"GENIUS":

(Gift or Curse?)

"GENIUS":
(Gift or Curse?)

Biological Origins, Key Modifiers,

Burdens, and Legacies

JAMES B. MACLEAN

ISBN: 1494245426
ISBN 13: 9781494245429
Library of Congress Control Number: 2013921719
CreateSpace Independent Publishing Platform
North Charleston, South Carolina

CONTENTS

v

CONTENTS

PROLOGUE

GENIUS AND EXTRAORDINARY COGNITIVE ABILITY (ECA) are not equivalent terms and have clear differentiating definitions that will be discussed in part I. Sources of genius and ECA, the crucial nature of very early environmental influences important in determining direction and focus, societal impact of the phenomenon in general over time and, specific individual engagements and legacies over the ages continue to attract a great deal of interest from multiple academic disciplines. Will and Ariel Durant, in their book *Rousseau and Revolution* (1967), focused on ECA key issues and genius as to its dramatic and unexpected appearance coupled with its potential legacy when discussing Jean-Jacques Rousseau in eighteenth-century Europe.

"How did it come about that a man born poor, losing his mother at birth and soon deserted by his father, afflicted with a painful and humiliating disease, left to wander for twelve years among alien cities and conflicting faiths,...suspected of crime and insanity, and seeing, in his last months, the apotheosis of his greatest enemy— how did it come about that this man, after his death, triumphed over Voltaire, revived religion, transformed education, elevated the morals of France, inspired the Romantic movement and the French Revolution, influenced the philosophy of Kant and Schopenhauer, the plays of Schiller, the novels of Goethe, the poems of Wordsworth, Byron, and Shelley, the socialism of Marx, the ethics of Tolstoy, and, altogether, had more effect on posterity than any other writer or thinker of the eighteenth century in which writers were more influential than they had ever been before? Here, if anywhere, the problem faces us: what is the role of genius in history?" (Durant 1967).

This ECA phenomenon or genius raises many questions as to its origins, the burdens, individual behavior, its impact on peers, and legacies for the future. What is the source of the underlying energy that drives the concentration and work ethic of geniuses and what determines the specific focuses they choose? Does early education play a role in the development of genius or is it simply a matter of genetics? Is there a common theme in terms of relationship with others? What is the degree of psychological discomfort, is it predictable, and is it necessary for the process of extraordinary cognitive ability or genius to become manifest? Psychological discomfort refers to clinically apparent mental disorders such as depression, acute and chronic anxiety, panic attacks or even intermittent brief psychotic episodes. How important is the manic-depressive personality trait for the phenomenon to mature? Where does the basic drive, motivation, or energy fundamental for the process come from? What may be identified as the primary trigger that stimulates the specific focus? What is the importance of specific peer exposures, the daily life routine, solitude, physical and mental health, and does one need to have clinical manic-depressive disease to operate successfully with ECA? What are the key modifiers, various environmental exposures that may significantly influence the direction and energy of intellectual pursuit, that seem so important at the time of early development and if identified, could they be modified or manipulated to promote positive direction for the basic driving cerebral energy that seems to be the core feature of these individuals? These and many other questions have attracted a great deal of interest from a variety of academic disciplines seeking to better understand the origin of higher cognition in man as well as the clinical manifestations and perhaps most importantly the specific legacies of the heightened intellectual ability.

As a clinical neurologist with a companion interest in biography and history, the biological origin of this unique faculty has held special interest for me, along with the remarkable individual legacies that some individuals with ECA have left for society. Less often addressed but equally fascinating, and perhaps of more practical importance to society in general, is the issue of accelerated cognition and the faculty of genius without clear direction or the more obvious and perhaps more destructive negatively directed focus with consequential societal destructive legacies. Clearly, a better understanding of the ECA phenomenon, its sources, and the consequential environmental modifiers is so important that determining its direction are of not only academic interest but also significant practical significance. A better understanding of the phenomenon could allow important early recognition and thus potential positive changes in the key modifiers known to be crucial for both the direction and intensity of this heightened intellectual energy. The potential modification could promote a more positive individual mental and physical health and a positive societal outcome— always an issue with this group. There is then an obvious unique but difficult challenge for the family and peers, along with the surrounding societal milieu, to not only recognize this special faculty early in life but also successfully influence it in a positive way in terms of the intellectual focus and the overall individual adaptation in terms of personal health and general integration into society.

These questions will be addressed to not only more clearly understand the origins of the phenomenon and its impacts on peers and future generations but also discuss potential gains from the early recognition of its existence and thus the potential for positive modification.

> "Of course in the son, it was controlled and magnified to an unimaginable extent, but the mystery of inheritance remains. And it is a mystery, how such gifts and qualities can be transmitted from generation to generation until there comes a time when in one particular person they blaze out and, in a sense, devour all those around."

> —Charles Dickens (Ackroyd 1990)

INTRODUCTION

WHAT EXPLAINS THE ASTONISHING MUSICAL COMPOSITIONAL accomplishments of Franz Schubert during his brief life, the remarkable written portraits of mid-nineteenth-century London by Charles Dickens, the more than a hundred symphonies of Joseph Haydn, and the remarkable scientific observations of Albert Einstein? How is it that clearly remarkable achievements such as these occur, and how can we explain what at first glance appears to be purely a random phenomenon? These initial musings have led me to develop this discussion, combining my lifelong interest in biography and history with a career in the clinical neurosciences.

"It was the best of times, it was the worst of times..."; "I went to the woods because I wished to live deliberately, to front only the essential facts of life..."; the *Eroica* symphony by Beethoven; Tchaikovsky Violin Concerto in D, op. 35; "Four score and seven years ago..."; $E=MC^2$; "We hold these truths to be self-evident..."; a Chopin prelude; Impressionist art; the *Mona Lisa*, and the theory of relativity all remind us of the extraordinary contributions of a few individuals to humankind's awareness of self and the world in which we live. The mystery of genius or extraordinary cognitive ability (ECA) has fascinated and puzzled many generations, with the dilemma clearly summarized by John Adams in eighteenth-century England, when he wrote about his distress as to "how little evidence remains of Shakespeare, either of the man or the miracle of his mind": "There is nothing preserved of this great 'genius'…which informs us as to what education, what company, what accident turned his mind to letter and drama" (McCullough 2001).

ECA continues to create considerable research interest from a wide variety of academic perspectives utilizing multiple discipline-specific study techniques. The neuroscientists are interested from a genetic, neurobiological, neurological, psychological, and social science standpoint. Others, including educators, philosophers, historians, biographers, behavioral scientists, and nonprofessionals, are equally intrigued with the mystery of the originating mechanisms of this phenomenon, its sustained energy, and the implications that this extraordinary human ability has for present and future generations. Is this extraordinary cognitive ability a gift or curse? It may be a gift for humanity in general but under closer examination, may be a curse for individuals who must deal with the intense inner tension that seems to always accompany their markedly accelerated thinking ability. "Genius," a lay term, has a variety of accepted definitions that generally imply achieving eminence through creating useful work with an enduring reputation because of the individual's extraordinary cognitive ability combined with hard work. Initially, it is important to differentiate these two terms from each other. ECA describes a basic biological process. Genius is a clinical term referring to certain individuals who achieve greatness by creating innovative but useful outcomes through individual efforts. Most would agree that ECA is the core asset in persons of genius initially endowed with a common but unique neurobiological template responsible for accelerated cognition, coupled with markedly advanced memory capacity, all of which may be significantly modified by specific environmental exposures, especially early in life.

This ECA phenomenon or genius as a unique clinical outcome, with its attendant extraordinary cognitive ability and creativity, raises many questions as to what accounts for the difference between the rare extraordinary human cognitive ability and much more common normal to heightened intellectual pursuits. As far back as the tenth century, notations of individuals with remarkable memories and intellect were attributed to "gifts of nature." Basic intellectual capacity of individual persons is as different as there are numbers of individuals studied. It is very clear, however, that certain subsets of the population have extraordinary intellectual capacity many levels beyond the highly intelligent person, and this capacity seems to date from very early in their development, if not from birth itself. Is there a single or common origin for ECA or genius, and if so, what is it? Is it multifactorial? If ECA is the basic core ingredient, does it always result in positive directions, and if not, what occurs when this unique energy is directed negatively, and, maybe more important , what are the consequences when the energy is not directed at all?

The legacies of this unique group have their own fascinations, with important influences on peers and future generations alike. Initial interest usually begins with biographic studies of specific individual behavior, performance, and creativity, with subsequent eminence well beyond the norm. History is replete with the extraordinary lives of people in many different pursuits who stand apart, identifiable even at a very young age, and who go on with immense intellectual capacity and ability to create with significant permanent impacts on their societies and future generations to come. What is the personal price that these individuals pay for this "gift," and is there commonality among them in this regard? Is the primary variation that accounts for the phenomenon an aberration in brain development, or does it manifest only under special environmental influences, or, more likely, is it a complicated combination of both?

What is this phenomenon called genius? The neurological, behavioral, and historical approaches focus on outcome and recognize creativity as the ultimate benchmark of genius. However, the basic scientist would argue that the genius phenomenon may occur without ultimate recognition or success because it is a basic anatomical, physiological, or neurochemical process that occurs and proceeds whether the individual is identified as being a genius or not. Measures are difficult and go well beyond the standard IQ test, which may be relevant to creativity. The most important measure may be simply nerve conduction velocity, cerebral energy, and memory storage capacity with accompanying intensity of thought. A clearer awareness of this process may allow us to better understand more standard mechanisms of cognition. Does genius always result in dramatic positive societal impacts or can genius lead to very negative societal outcomes or even more concerning, because of frequency, clinical behavior that is adversely changed by the underlying ECA without positive or negative impacts to society but significant impacts on individuals and their interaction with society?

My intent is to thoroughly study this ECA phenomenon from four primary perspectives. These include:

1) Discussion of the basic neurobiological framework necessary for its manifestations with the various biological mechanisms understood to be important for memory, perception, and cognition

2) Clinical observations of the multiple personality and behavioral features of these individuals

3) Multiple comparative biographic studies

4) Speculations about individuals with ECA without clear direction or negatively directed societal influences as to their individual struggles, the origins of the specific focus or lack thereof, and their impact on the surrounding milieu.

Important legacies of critical contributions to humankind's understanding over the years because of this phenomenon will be briefly mentioned. The legacies of highly creative people with this unique phenomenon are well known, but the much larger potential impact of individuals with this unique capacity but for various reasons are not recognized or fail to become creative in a positive manner or even more important convert this energy in negative societal directions is not addressed or as well understood. With a more complete understanding of the potential causes of this unique human facility of ECA, we may well be able to identify these individuals earlier and perhaps modify some key environmental modifiers leading to a higher probability of a positive outcome.

The goal of the work is to better understand the phenomenon of genius from many different perspectives, including core neurological framework, clinical behavioral studies, and individual biographies while detailing the legacy, both positive and negative, of this unique facility.

PART I:
HISTORICAL REVIEW OF "GENIUS," DEFINITIONS, KEY QUOTATIONS, AND LEGACIES:

"Under the influence of congestion of the head, many persons become poets, prophets and sibyls, and, like Marctis the Syracusan, are pretty good poets while they are maniacal; but when cured can no longer write verse."

—Aristotle

A. HISTORICAL REVIEW OF THE ORIGINS OF ECA AND GENIUS

HISTORICALLY, THERE HAVE BEEN MANY widely variable but interesting theories as to the origins of genius and what makes up the underlying critical criteria necessary to identify a person of genius. The time-honored debate, nature versus nurture, continues as to whether geniuses are born or develop from critical environmental influences or if the phenomenon depends on a very delicate balance between the two. Theories from antiquity can be traced to initial Greco-Roman speculation that believed the source of genius to be the divine, from the muses, or specifically occurring from a significant melancholic personality that was responsible for its manifestation. Cesare Lombroso connected genius to mental illness and identified epilepsy, melancholia, megalomania, alcoholism, moral insanity, and syphilitic paresis as common to the highly creative (Lombroso 1891). As he said, "Men of genius have an exquisite and sometimes perverted sensibility with depression and paranoia." Lombroso also thought that genius was a degenerative psychosis of the epileptic group. The convergence between madness and genius was also discussed by Theophilus Hyslop (Britain)

(Steptoe 1998). Schopenhauer, according to Lombroso, thought that there was a very close relationship between insanity and genius. There have been many early theories of divine intervention as the specific origin of this phenomenon. The heightened intellectual behavior was implanted by God as the individual was chosen as a direct source by God to utilize these special intellectual facilities in a favorable fashion for the general good. There also were many theories relating genius to madness early on with only a very slim line between the two. Some developed theories that genius might be a form of psychosis and certainly severe melancholy was quite common, with some postulating that depression is the essential key element in all persons of genius. Both manic and other affective elements, including schizophrenic features, may be relevant to creativity (Runco 2004 and Kiehl 2006). More recent theories relate to an occurrence of a unique nonmedelian alignment of neuronal-directing genes occurring by chance alone in very early brain development. This is labeled emergenesis, which will be discussed later in more detail. Lykken, at the University of Minnesota, has done some more recent, important genetic work related to higher cognitive ability (Steptoe 1998). Doubtless, as time moves on, many new neurobiological theories will be developed as to the origin of genius as more modern brain-scanning techniques are perfected and genetic engineering and attempts of cloning of genius occur.

> "A seething cauldron of ideas, where everything is fizzling and bub-
> bling about in a state of bewildering activity, where partnerships can
> be joined or loosened in an instant, is genius."
> —William James (Simonton 1999)

B. DEFINITIONS COMPARING THE NEUROBIOLOGICAL AND LAY DEFINITIONS OF ECA AND GENIUS

HUMANKIND HAS BEEN INTRIGUED FOR CENTURIES by this phenomenon called genius as to its origins, its clear uniqueness with its remarkable separation from the norm, the apparent early age of onset, individual interactions with peers and society, and profound individual legacies for future generations. This discussion begins with clear definitions as to what the term *genius* means and what it doesn't and how the term is different from extraordinary cognitive ability (ECA). The clinical term *genius* typically implies a level of functioning beyond the ordinary, with characteristics of high energy in general, intensity in all pursuits, marked creativity, high intellectual powers, independence, restlessness with unhappiness, great self-imposed sacrifice, daily heavy psychological burdens, huge desire for solitude, and very little patience for the day-to-day mundane tasks of life. Genius is the ability to understand and communicate universal truths. Genius is the proven ability to produce artistic, scientific, or other intellectual work that is considered supremely valuable during or after the lifetime of the producer, according to Hershman and Lieb (1988). This clinical term *genius,* with the associated behavior described, needs to be differentiated from ECA, which is a neurobiological term defining a specific human attribute of unique thinking ability that results from a certain constellation of basic neurobiological elements.

The term *genius* comes from the etymological root *gigno,* meaning "I bring forth," "exceptional talent," or "exceptional creativity." ECA describes a unique human facility that occurs with the development of a rare but specific neurobiological template, to be discussed in the later basic science discussion, representing anatomical, chemical, and physiological changes early in brain development that in combination are responsible for the appearance of markedly accelerated higher cognition, some features of which may even be apparent at birth. Most of these structural changes occur prior to birth and thus provide the basic ingredients necessary for accelerated cognition. This term therefore describes the potential for this unique human cognitive ability that occurs with the development of these remarkable brain operative parameters that provides for the manifestations of markedly accelerated cognitive ability. Genius, on the other hand, is a lay term that describes individuals with obvious accelerated cognitive ability and who obtain extraordinary human achievement that may be in large part due to the facility of ECA. ECA implies an advanced brain mechanism, while genius is what occurs when this mechanism is used to its greatest potential. *Genius* thus describes one who attains eminence by leaving for peers and posterity alike an impressive body of contributions that are both highly creative and useful (Rochenberg 1976). This may be one result of this unique underlying anatomy and physiology, but the basic mechanism of higher cognitive ability has an impact well beyond this narrow definition. Perhaps my study is not so much about genius as it is about the underlying cerebral energy in place that allows the phenomenon of genius and creates other issues in human culture that do not fit this definition. I think that geniuses or persons with ECA who are creative both understand and are able to communicate universal truths that we all will eventually recognize as being correct once they are presented to us. Emerson's quote below describes this very well. An example of which are the wisdoms of Thoreau and Emerson detailed in their writings as to the relationships between humankind and nature that when presented represents recognizable universal truths.

"Creativity" has always been accepted as the key ingredient for genius and many have devised definitions of what we should mean by this term. Carl Roger's definition of creativity involves developing a novel construction or product, see Carl Rogers Quotes. This novelty grows out of the unique qualities of the individual in his or her interaction with the materials of experience (Rochenberg and Hausman 1976). Creativity is also commonly defined as the ability to produce work that is novel, useful, and generative (Fink et al. 2007). Creativity is the ability to find unity in what appears to be diversity. Therefore, creativity is the ability to unite diverse elements and display order. It is the expression in a systematic fashion novel to unite orderly relationships (Heilman 2009). Webster describes genius as "extraordinary...intellectual power especially as manifested in unusual capacity for creative activity of any kind." Webster's New Collegiate Dictionary second edition 1953.

ECA, in contrast, is biologically descriptive of an extraordinary thinking ability that may be crucial for creativity unique enough to be labeled as a work of genius. Creative genius has been separated or differentiated from eminent persons in noncreative occupations (Rothenberg and Wyshak 2004). One could argue that eminent persons in noncreative pursuits could still be labeled as genius, although most would believe that remarkable creativity is a core attribute of this phenomenon. It is also clear that many individuals with ECA are never identified as highly creative and thus are not called genius but have the same unique neurobiological template for ECA that persons of genius have. This differentiation introduces the very important issue of how to classify individuals with ECA, who never focus and do not achieve remarkable contributions and thus are not thought about as geniuses. These individuals nevertheless must navigate many of the same potential burdensome features that persons of genius must deal with. This will be further discussed in part VI.

These two core terms, genius and ECA, have originated from separate approaches to understanding accelerated thinking. *Genius* is the much older term and has been used historically to label persons who have displayed a unique and remarkably accelerated level of human thinking or cognition well beyond those persons thought to be at the high end of the cognitive ability curve. However, an important second criterion for the definition of genius drives the clear need to separate the term *genius* from ECA. This is that a genius must not only be exceptionally intelligent but also achieve a degree of eminence resulting from useful contributions from work.

These two terms, *ECA* and *genius*, will be used throughout this discussion but are not to be interpreted as having the same meaning for all of the reasons discussed above. It also should be clear that all persons with ECA do not necessarily become geniuses, but all geniuses most likely require ECA as the core asset necessary to achieve genius. There are some very interesting direct quotes through the ages as to what genius is, most of which have come from persons who most likely were geniuses in their own rights. No more insightful one may be Emerson's from his essay on self-reliance, where he focuses on what may be the core ingredient that first tips the hand identifying a person of genius:

"Great works of art have no more affecting lesson for us than this. They teach us to abide by our spontaneous impression with good-humored inflexibility than most when the whole cry of voices is on the other side. Else, tomorrow a stranger will say with masterly good sense precisely what we have thought and felt all the time, and we shall be forced to take with shame our own opinion from another."
—Ralph Waldo Emerson from "Self-Reliance"

To believe your own thought; that is genius."
—Ralph Waldo Emerson (Richardson 1995)*

"The Main image of the creative mind is of a volcano."
—Ralph Waldo Emerson (Richardson 1995)

"Genius is the activity that repairs the decay of things."
—Ralph Waldo Emerson (Richardson 1995)

"There is no history, only biography."
—Ralph Waldo Emerson (Richardson 1995)

"The true genius is a mind of large powers, accidentally determined to some direction."

—Samuel Johnson

"Genius must be born, and never can be taught."

—John Dryden

"Civilizations are often defined by the lives and works of their creative geniuses."

—Dean Keith Simonton (Simonton 1999)

"Extreme intelligence was very near to extreme madness."

—Pascal

"I have always been regarded as a man specially favored by fortune… But…I might go so far as to say that in seventy-five years I have not known four weeks of genuine ease of mind."

—Johann Wolfgang Von Goethe

"Chopin was…of an intensely passionate, an overflowing nature…. Every morning he began anew the difficult task of imposing silence upon his raging anger, his white hot hate, his boundless love, his throbbing pain, and his feverish excitement, and to keep it in suspense by a sort of spiritual ecstasy—an ecstasy into which he plunged in order to…find a painful happiness."

—Franz Liszt describes Chopin (Hershman and Lieb 1998)

"The future, in its justice, will number him among those men whom passions and an excess of activity have condemned to unhappiness, through the gift of genius."

—Delacroix speaking of his friend, Chopin (Szulc 1998)

"I never let my schooling interfere with my education."

—Mark Twain

C: SAMPLE QUOTES FROM PERSONS OF GENIUS

IT HAS BEEN SAID THAT WISDOM expressed by persons with ECA often express truisms that we all innately recognize but have not been able to express ourselves.

Francis Joseph Haydn, truly a genius himself by all the criteria mentioned, identifies Mozart, a peer, as a genius, with some additional insight as to the potential fates of many persons of genius throughout the ages:

> "Ah, if only I could persuade every friend of music, but especially the great ones, to understand and to feel Mozart's inimitable works as deeply as I do and to study them with as great feeling and musical understanding as I give to them. If I could, how the cities would compete to possess such peerlessness within their walls. Prague would do well to keep a firm grip upon this wonderful man—but also to reward him with treasures. For unless they are rewarded, the life of great geniuses is sorrowful and, alas, affords little encouragement to posterity to strive more nobly; for that reason so many

promising sprits succumb...it angers me that this unique man Mozart has not yet been engaged by some imperial or royal court. Forgive me, honored sirs, for digressing, but I like the man too well." (Jacob 1950)

"Many a genius is spoiled prematurely by earning his bread by miserable work."

—Franz Joseph Haydn (Jacob1950)

"I don't measure a man's success by how high he climbs but by how he bounces when he hits bottom."

—General George Patton

In the following quote, Charles Darwin addresses perhaps the key essentials common in persons with ECA: highly focused intellectual energy combined with a keen curiosity:

"I think that I am superior to the common run of men in noticing things which easily escape attention."

—Charles Darwin (Simonton 1999)

Albert Einstein speaks to the need for isolation and independent thinking so crucial for original creativity:

"Such isolation is sometimes bitter but I do not regret being cut off from the understanding and sympathy of other men.... I am compensated for it in being rendered independent of the customs, opinions, and prejudices of others and am not tempted to rest my peace of mind upon shifting foundations."

—Albert Einstein (Gardner 1993)

"Life is like riding a bicycle. To keep your balance you must keep moving."

—Albert Einstein, in a letter to his son Eduard, February 5, 1930

Leo Tolstoy expresses wisdom as to the prime importance of intentions as opposed to contemporary deeds, which may be of great value as judged by the ages:

"I think that we will be judged by our conscience and by God, not for the results of our deeds which we cannot know, but for our intentions, and I hope that my intentions were not bad."

—Leo Tolstoy (Noyes 1918)

"I don't know what I may seem to the world, but to myself I seem to have been only like a boy playing on the seashore and diverting myself now and then in finding a smoother pebble or a prettier shell than ordinary, while the great ocean of truth all undiscovered before me."

—Sir Isaac Newton, per George Combe, 1828

D. THE LEGACIES OF ECA AND GENIUS

HUMAN HISTORY IS REPLETE WITH A multitude of examples of human achievement with major legacies that in large part occurred secondary to ECA. From Bach to Michelangelo, Leonardo da Vinci to Chopin, Mozart and Beethoven to Emerson, Dickens to Newton, Einstein to Emily Dickenson, and more contemporary individuals, such as Churchill and Gandhi—these individuals represent multiple genres, including art, literature, science, and leadership, to only name a few. ECA does not confine itself to any limited genre and may be expressed in any focus of human behavior, including negatively directed endeavors, such as those of Hitler, Napoleon, Stalin, and perhaps the more contemporary Syria's Assad and Iraq's Saddam Hussein. While these individuals most probably have ECA, they are not geniuses and do not fit the clinical definition of genius that includes novel and useful contributions to humanity. It is not necessary here to point out the numerous individuals throughout history meeting the criteria pointed out in section IB for genius. But it is best to consider the broader issue of its origins, the major influencing environmental factors, the potential benefits of early recognition and what that might imply, and the emotional and physical price they all seem to pay.

The issue as to what motivates them or is responsible for the persistent, extraordinary drive crucial for their individual successes remains a mystery be it nature or nurture or a delicate combination of the two. The outcome of this exceptional intellectual ability appears to be quite sensitive to the multitude of potential environmental exposures, which may well be responsible for the direction of the intellect and its intensity in these individuals. One of the more intriguing elements of all this in terms of legacy impacts of ECA is the apparent random choices of focus that these individuals pick. These may be highly positive for contemporary and future generations or just the opposite, with a destructive impact. However, the variations are not random, as key early environmental modifiers may now be understood to have great influences on these individuals and the directions their respective intellects take them.

On balance, the human condition has greatly benefited throughout history from this ECA phenomenon and genius, with the multitude of advancements in many fields of human endeavor that has made life much more enjoyable.

PART II:
JOHN DOE AS
FICTIONAL BIOGRAPHICAL
EXAMPLE OF ECA

"There is no doubt that, in the lives of writers, the shadows of a grandfather or grandmother (most important even when they are not clearly discerned) can be seen lying across the paths they follow. It is as if the peculiar chemistry of genius sometimes skips a generation, as if it is the nature of the grandparents that really accounts for the temperament and even behavior of the one who comes after."

—Charles Dickens (Ackroyd 1990).

A: INTRODUCTION

As an introduction to clearly understanding ECA origins, its clinical manifestations, and its overall impact on the human condition, an initial presentation of a fictional biography of John Doe, scripted in a modern context, may provide a useful starting point. Multiple key features are common and important in the development of this unique ability and are introduced in this fictional biography. The biography can be referred to as later sections address our present understanding of this process..

This fictional biography highlights many common clinical manifestations and displays many issues that have both basic and clinical science implications that may help us better understand extraordinary cognitive ability origins.

B: JOHN DOE, CLINICAL SCIENTIST

"I believe in the Omni-presence; that is, that all is in each particle; that entire Nature reappears in every leaf and moss. I believe in eternity—that is that I can find Greece and Palestine and Italy and England and the islands—the genius and creative principle of each and all eras in my mind."

—Ralph Waldo Emerson (Richardson1995)

INTRODUCTION

John Doe continues to work daily on the frontiers of medically related genetic research. In his relatively young career, he has identified several significant gene loci modifications responsible for degenerative neurological disorders, but perhaps more importantly, he is now in the process of developing an all-encompassing model or theory for human inheritance that may explain our unique vulnerability or lack thereof to many medical conditions. He has published his research and is considered by his professional peers to be on the cutting edge of human genetic research, although many of his contemporaries are not convinced that his most recent theories of genetic code transfer and human illness susceptibility will lead to the practical revelations and clinical usefulness that he is predicting.

John Doe's family members have been tradespeople, teachers, and small-business owners in rural Virginia for decades. He was born in Charlottesville, Virginia, in 1970, the second son of Mr. and Mrs. Matthew Doe. His mother was raised in Virginia, and his father grew up in South Carolina. His family dates back several generations, residing primarily in the southeastern United States. It is interesting that he has no known family members in either science or the medical field, although he did have a great uncle who practiced general medicine in a small town in West Virginia. Other family members have attended college over the years with subsequent careers in business and teaching, but no individuals in the family tree analysis had clear ECA or were thought to be especially noteworthy in terms of personal attainments.

His story is of a unique interest and focus, without any obvious preceding models or predictors that launched in an altogether new direction with his life and interest, obtaining impressive academic credentials along the way but primarily focusing on unique scientific questions that most of his colleagues and peers had difficulty understanding. To date, his accomplishments are clear, with the identification of several gene modifications that help explain the appearance of rare human neurological disorders. This has elevated him to an elite status among colleagues in genetic research. Many of his academic colleagues now, however, think that his efforts in developing an all-encompassing hypothesis for inheritance and human medical conditions is much beyond our present state of knowledge of genetic methodology, but many admit that they have a great deal of trouble understanding his formulations.

Given his intensity of focus and his developing genetic model hypothesis, he may well be considered in the future as one of the true giants in the field of medical genetics because he developed important theories related to disease origin and long-term medical therapeutics.

BIOGRAPHY

For many generations, John Doe's family has lived and worked in a relatively small geographic area of the southeastern United States. It is not clear when distant preceding generations initially immigrated to the Americas, but we have some information of great-grandparents on the maternal side residing in West Virginia. There are no apparent members of the family from recent or distant generations who were particularly academically inclined or achieved extraordinary intellectual accomplishments in their lives. His father was raised in Raleigh, South Carolina; attended school in South Carolina and Virginia; and met his wife in Charlottesville, Virginia, where she was obtaining a teaching certificate at the University of Virginia. They made their home on a small rural farm near Charlottesville as his father developed, owned, and managed a small retail business in the city. She became a well-respected high school teacher of English literature.

John Doe was born in January 1970, the second child of Mr. and Mrs. Matthew Doe. He had an older brother and two younger sisters. He was clearly quite different from his siblings, both in terms of temperament from a very early age if not from infancy. He never became the "norm" as judged

25

by the personalities of his three siblings and peers. He was a difficult child to raise and seemed to be always hyperirritable. He would admit that he felt quite tense most of the time, with his mind moving in multiple directions at the same time. He had a great deal of difficulty relaxing and enjoying conversations with others. He always had difficulty listening to others, as he rapidly became impatient, primarily from "boredom." He did not relate well to any members of the family, especially his older brother and father, perhaps because of a much more focused interest in personal pursuits as opposed to group or shared activities. He displayed a unique curiosity for all kinds of things, including the environment, from a very early age. He was always asking many questions but never being comfortable or satisfied with the answers that he was receiving. He loved spending hours focusing on the natural environment of the farm and surrounding countryside and soon became a prodigious reader.

His early education in the rural school system of Virginia did not go well. He lost interest in schoolwork very quickly and needed to be prodded continually to get his school work done. He made very few, if any, real friends at school or in the neighborhood, seemed impatient with peers, and was not interested in the usual activities of children and young adolescents. He was a loner in high school, without much interaction with others. But he somehow passed through the grades without too much difficulty, despite the fact that he did not seem to apply himself or focus to any extent. Test scores were reasonable to quite good with very little perceived effort.

He spent most of his time by himself, absorbed in books or wandering around the countryside. His relationship with his father proved to be very problematic. Over the years, his father could never understand him and found it difficult to lead him in any direction.

After graduating from high school, he wanted to attend college and major in biology, primarily because of his keen interest in the natural sciences. He did this at a small community college in Richmond, Virginia. Routine academic work at this school was difficult for him to pursue, but several of his professors recognized an unusually quick mind and wit with his avid interest in basic science and thus directed him into individual, creative projects in basic science. He later returned to Charlottesville after being admitted to the medical school based on his unusually high MCAPS scores and very favorable recommendations by his teachers at the community college in Richmond.

Medical school was very difficult for him, as he much preferred to spend his time on individual pursuits in scientific research. While in medical school, he developed a keen interest in human genetics and participated in several summer projects at the school. The first two years of medical school were much more interesting for him than the later clinical years. He found that he was much happier understanding scientific puzzles and developing his own research projects based on the questions that he had than he was practicing clinical medicine, as his great uncle did.

After his internship in Charlottesville, he joined a fellowship program in medical genetics at Columbia Presbyterian Medical Center in New York City, where he studied with John Smith, a pioneer in the initial investigation and understanding the human genetic code. He assisted Dr. Smith with several projects and was instrumental in discovering very important gene modifications responsible for causing several neurological degenerative diseases, the causes of which were not previously understood.

With his achievements at Columbia, a Dr. Jones at Harvard, who had just procured a large grant to extend his genetic studies, invited him to work with him in his lab of medical genetics and help with teaching commitments that he had both at Harvard College and the Harvard Medical School. John Doe decided to do that and joined the Harvard staff and Dr. Jones in his laboratory five years ago. He does a modest amount of teaching, which is required of him as a member of the medical school's department of genetics. However, he spends almost all his working hours in the laboratory, working on the genetic code and his revolutionary human gene transmission theories. He has no leadership or administrative duties at the university, and for the most part, he works alone with a few laboratory assistants. He occasionally attends national medical meetings that focus on genetics. This often proves frustrating for him as he has difficultly accepting the level of academic conversation and prefers to spend his working hours outside the lab writing his research about new gene identification and developing his all-inclusive theories of individual unique human genetic inheritance and its importance as the primary determinant of individual illness threshold throughout life. He has very few social acquaintances and spends most of his time away from his work, reading prodigiously and occasionally attending cultural events in Boston.

At age forty-three, he continues with a very active career in medical genetics. He was married briefly about ten years ago and still visits his mother in Virginia several times a year. He has found it very difficult to relax and has always thought that he had to engage with a high sense of urgency and energy to keep from becoming depressed. He has not had severe clinical depression but certainly has had clear features of a manic-depressive disorder. His general health had been good, but he felt that he was beginning to "slow" down, lacking the energy that he once had. He has not had any desire to advance in the academic community other than to be able to work his own projects in the genetic lab and continue to develop his unique theories of individual human thresholds to various acquired and inherited medical conditions as transmitted by our unique genetic code on the genome.

He has been and remains highly regarded in the academic community, especially for discovering several significant and specific gene variations responsible for two neurodegenerative disorders, the pathogenesis of which previously have not been understood. His colleagues are intrigued by his new theories relating individual human disease thresholds to specific inherited DNA patterns, but most believe that he is ahead of the known science at present and most have difficultly following his ideas.

John Doe and Accelerated Cognitive Ability

At a very early age, John Doe had multiple features that signaled ECA. He was highly energized and had a unique curiosity but found it difficult to relax or enjoy the more common pursuits of his peers. He had a keen wit but a great difficulty in social settings, having very little patience with family and peers alike. He did well in school without apparent effort and devoted most of his time to individual pursuits, such as prodigious reading and discovery. He has been able to apply considerable focus to his academic work in the field of human genetics and is in the process of leaving a significant legacy of new knowledge for future academicians in the field. He prefers his own companionship to that of others, as he likes to think and pursue at his own speed and finds that he has very little patience depending on others. All these attributes are indicative of ECA.

JOHN DOE: FEATURES OF COMMONALITY AND DISSIMILARITY WITH OTHERS WITH ACCELERATED COGNITIVE ABILITY

As mentioned above, John Doe has many of the common personality traits and features of individuals with ECA. Features of dissimilarity with others with ECA include a seemingly more controlled inner tension without significant medical illness or serious psychiatric issues that seem quite common in individuals with ECA. He remains quite young at age forty-three, so it is difficult to identify potential areas of dissimilarity yet to manifest.

JOHN DOE AND HIS LEGACY

His legacy to future generations and specifically to medical genetics remains speculative, but he has already helped identify the genetic origins of several neurodegenerative disorders that for years had remained a puzzle as to causation. Perhaps more important, his new concepts about the individual human susceptibility to various acquired and degenerative disorders determined by unique genetic DNA configurations and its implication to future clinical diagnostics and therapeutics may be considered a giant leap forward in medical science.

PART III:
JOHN DOE MODEL
BIOGRAPHY AND THE
NEUROBIOLOGICAL
FRAMEWORK FOR
COGNITION

"Genius is the faculty of seizing and turning to account everything that strikes us."

—Johann Wolfgang Von Goethe (Richardson 1995)

A: JOHN DOE BIOGRAPHY WITH DISPLAY OF BASIC THEORIES AS TO ECA ORIGINS

THE FICTIONAL BIOGRAPHY OF JOHN DOE presented in part II, above, serves as a modern biographical example of this unique "gift" of heightened intellectual ability to introduce the multiple factors involved for the appearance of this phenomenon. It combines the basic neurobiological theories of origin with the recognized important environmental markers or modifiers that are crucial for individual outcomes. This fictional biography is purposefully full of many of the key elements thought to be crucial for the appearance and course of the phenomenon in these individuals. This is displayed with a relative lack of family members from previous generations equipped with this heightened intellectual curiosity. In addition, many behavioral attributes and important exposure to the influence of key modifiers early in his life had major influence as to his focus and personal adaption to the surrounding milieu, including features of personal and emotional health. As we will see when we review multiple individual biographies of individuals with ECA, a great deal of commonality exists in persons who have extraordinary cognition, not only in terms of their familial backgrounds but also in exposure to various important modifiers early in life, subsequent behavior, and in physical and emotional health in general. Initially, the exploration of the basic neurobiology provides the source in some detail as best we know it at present, including a specific non-Mendelian genetic theory of origin along with an expanded examination of the critical early modifiers so important in the overall outcome of the phenomenon.

"So much of our time is preparation, so much is routine, and so much retrospect, that the pith of each man's genius contracts itself to a few hours."

—Ralph Waldo Emerson (Richardson 1995)

B: BASIC SCIENCE SECTION

1. INTRODUCTION

Understanding the phenomena of extraordinary cognitive functioning must begin with an extensive review of the multiple and varied neurobiological processes presently understood to be important for higher cognition. These involve an array of biological issues, including genetic predisposition, early brain development with specific anatomical structures forming and being coupled with variable chemical and physiological pathway developments that ultimately form the biological template for cognitive functioning. Clearly, with preliminary observations, distinct variability in the level of cognitive ability with capability ranging from the cognitively handicapped to those individuals with obvious extraordinary higher cognitive ability exists. The basic macro biologic framework for cerebral functioning is constant, but the crucial variability is determined by the apparent infinite variations in microscopic detail in genetic design, anatomy, physiology, chemistry, and degree and intensity of neuroplasticity (neuronal wiring modifications over time) with early development. Thus, the study of the potential extraordinary cognitive ability at birth must begin with understanding the basic biological determinants of cognition, attention, and memory. These are initially designed by unique combinations of the DNA that make up the genome (genetic code framework) and are potentially modified in a complex, variable manner with very early brain developmental process. Do we sufficiently understand the basic mechanisms involved in attention, memory, and cognition to be able to identify important variations that might account for different levels of cognitive ability? Assuming that we do, one then can go on to speculate as to the source of higher cognitive ability by examining in detail various aspects of the overall process of cognition. This discussion will examine the areas of interest from a basic neuroscience template perspective, including its originating genetic design, which may be important for the obvious degree of variation that then becomes modified by key environmental factors. This modification leads to focusing on two remaining sections of this discussion, namely, clinical considerations of cognition and biographies of persons with accelerated cognition. As mentioned, critical genetic issues will be reviewed along with a discussion of the "emergenesis" theory, which details a unique genetic mechanism that may be critical for triggering the unique biological framework necessary for extraordinary cognitive ability. It will discuss and review this neurobiological template and the important anatomical, chemical, and physiological understandings that now form the underpinning of cognition, memory, and emotion.

2. DATA

Theories as to the origin of extraordinary cognitive ability must be consistent and integrated with our contemporary understanding of human brain anatomy, chemistry, and physiology. It is well accepted that specific circuits (neuronal connections) are involved for unique aspects of higher cerebral functioning. These same circuits process information to yield noncreative combinations and highly creative formulations. Modern cerebral research defines cognitive function as hierarchically ordered. The prefrontal cortex represents the top of the hierarchical anatomical framework for higher cognitive functioning. Cognitive abilities such as memory, attention, flexibility, prioritizing, multiple directional simultaneous thinking patterns, and velocity of thinking are all ascribed to the prefrontal cortex. To more thoroughly understand the mechanism of accelerated cognitive ability, primarily originating in the prefrontal cortex, the following items must be understood:

- Basic anatomical, physiological, and chemical makeup of this region, integrated with multiple other reciprocating brain regions
- The initiating genetic mechanism responsible for the basic anatomical design
- The multivariable, complex chemical and physiological processes that make up cerebral functioning help further identify potential variations that may account for observed differences in levels of human cognition

Key elements involve anatomical considerations of multiple patterns of neuronal hardwiring, available neurochemical transmitter concentrations and locations, specific synaptic formations (junction between neurons,) variable neuronal conduction velocities, the neuroplasticity of structure, and function with usage over time (Mesulam 2000).

GENETIC CONSIDERATIONS

Creativity or genius, at least in part resulting from ECA, has two defining criteria previously discussed namely the ability to produce work that is both novel and appropriate. Any theory of creativity or genius which are the core attributes of ECA must be consistent and integrated with the contemporary understanding of genetic mechanisms responsible for the design of the basic brain framework or the "neurobiological template." It also includes the vast amount of information now available about prenatal anatomical, physiological, and chemical brain mechanisms, all of which are critical variable features in any discussion about the origin and character of human cognition and creativity.

A comprehensive discussion of the basic science elements essential for ECA must begin with an understanding of or at least a general familiarity with what we now know about the genetic mechanisms responsible for the variable cerebral (brain) biological template and functionality of the operating prefrontal cortex. Sequencing the human genome has made it possible to begin identifying which of the many genes are critical to the central nervous system (CNS) formations that have an impact on human cognition (Goldberg and Weinberger 2004). One of the many intriguing aspects of this ECA discussion is the startling appearance of cognitively gifted persons in families where it is very difficult if not impossible to identify any preceding family members with similar cognitive traits or abilities, even after exhaustive genealogical reviews. This becomes very apparent with multiple biographical reviews of persons with accelerated cognitive functioning, where preceding family members do not appear to have similar focuses or the intense intellectual focus from early ages that seem so apparent in these select few gifted individuals. This intellectual energy, of course, has significant environmental implications, but more likely is driven by specific genetic patterns occurring with unique gene configurations that occur by chance and are not transmissible at least as understood in the classical genetic Mendelian tradition. This unique genetic inheritance known as emergenesis was originally conceptualized by D. T. Lykken (1982). It proposes by chance a unique gene alignment or polygenic configuration of DNA responsible for the key basic cerebral elements that are responsible for the accelerated cognitive ability appearing in a family without existing identifiable similar traits (Li 1987; Gardner 1988; Lykken 1982, 1992, 2006). Specific gene combination at multiple crucial genetic loci determines a phenotype. It designs the basic biological framework responsible for unique cognitive abilities but cannot be transmitted in the classical Mendelian sense, given the unique chance occurrence of this alignment as discussed by Lykken, Li and Gardner, referenced above. When a phenotype is determined by a specific gene combination of several loci, it is called an emergenic trait. Such a trait, although genetically controlled, does not run in families, because the specific gene combination cannot be preserved with subsequent reproductions.

With the DNA alignment in place at inception, determining the unique neurobiological framework of the developing brain, with its myriad of potential variations in structure and function, the human brain begins to develop with succeeding cell divisions. This leads to the embryonic and eventual newborn brain that was designed from the specific genetic code present with the very first cell division. Next, a complex interplay develops between inherited traits and environmental factors, in which genetics may underpin exposure to nurturing social and physical experiences.

With each unique genetic configuration, the very first cell division establishes, specific anatomical, physiological, and chemical formations occur that will make up the neurobiological cerebral template responsible for cognitive functioning.

ANATOMICAL FRAMEWORK

To more thoroughly understand the mechanism of accelerated cognitive function, the basic anatomical, physiological, and chemical makeup of a region important for thinking and integrated with multiple reciprocating brain regions must be understood in detail. The neuroanatomy of creativity is extensively discussed by Jung 2009 and Jung 2010.

The human brain displays marked regional variations in architecture, connectivity, neurochemistry, and physiology. The core of this discussion is to relate brain structure to behavior. Structural foundations of cognitive and behavioral domains take the form of partially overlapping large-scale networks organized around reciprocally interconnected cortical epicenters (Mesulam 2000). The brain is composed of several cell types, with the neuron being the primary anatomical element responsible for cerebral functioning. All the cells are interconnected by a myriad of network configurations responsible for cerebration. Two other cells are represented in large numbers in the human brain: the astrocyte, which has been considered as a supportive cell for the neuron, and the oligodendrocyte, which also is thought to be supportive and not a primary functioning cell in the same manner as the neuron.

What is the role of glia, a nonneuronal cell, and the cognitive process (Fields, 2004)? Interestingly, over the years, given the considerable interest in the origin of creativity or "genius," postmortem anatomical examinations have been done on some persons of "genius." One exam involved Albert Einstein's brain, and it was thought that his brain showed an "excessive" number of astroglial cells, which raises the speculation that perhaps for him, these cells were more than just supportive in nature (Colombo et al. 2006; Chklovskii, Mel, and Svoboda 2004).

Key cerebral elements to consider in attempting to understand variable cognitive ability involves anatomical considerations of multiple patterns of neuronal hard wiring, available transmitter concentrations, and at various specific locations, specific synaptic formations, variable neuronal conduction velocities, and plasticity (changes) of structure and function with usage over time. Memory storage in the human brain remains a mystery; however, a large amount of information is now available that suggests changes in synaptic morphology, protein synthesis, and gene expression are important sites of potential alterations that may be crucial for lifelong memory (Mesulam 2000).

The human brain consists of 10^{11} neurons connected by 10^{15} synapses. This awesome network has a remarkable capacity to translate experiences into vast numbers of memories and provides the substrate for thinking and creativity (Chklovskii, Mel, and Svoboda 2004). Initially looking at gross anatomical considerations, Flaherty 2005 presented a three-factor anatomical model of human idea generation and creative drive, focusing on interactions between the temporal lobes, frontal lobes, and the limbic regions of the brain. The focused aspect of creative arousal and its high goal-directedness may be driven from the middle anatomical region of the limbic lobe in combination with a high concentration of the neurotransmitter dopamine in that area. Her method of study included functional imaging or real time radiographic visualization during elevated cognitive activity drug studies, and post neuroanatomical damage analysis. She has argued that connections between the frontal and temporal lobes are more important than those between left and right hemisphere, and the addition of limbic system input is crucial for generating the creative drive. Right and left hemisphere lesions rarely affect creativity selectively alone in isolation (Flaherty 2005). A substantial amount of evidence exists that indicates specific, discrete neuronal circuits are dedicated to higher cerebral functioning or thinking. A basic assumption of the framework responsible is that neural circuits that process specific information to yield noncreative formulations of information are the same neural circuits that generate creative or novel combinations of information (Fink et al. 2008). Modern brain research conceptualizes cognitive function as hierarchically ordered. The cerebral cortex, and in particular the prefrontal cortex, is at the top of that hierarchy, representing the neural basis of higher cognitive functions. Creativity requires cognitive abilities, such as working memory, sustained attention, cognitive flexibility, and judgment of propriety, which are typically ascribed to the prefrontal cortex. Neural computation that generates novelty can occur during two modes of thought (deliberate and spontaneous) and two types of information (emotional and cognitive). Regardless of how novelty is generated initially, circuits in the prefrontal cortex perform the computation that transforms the novelty into creative behavior. To that end, prefrontal circuits are involved in making novelty fully conscious, evaluating its appropriateness and ultimately implementing its creative expression.

Creativity requires the novel understanding and expression of orderly relationships, and novelty requires that the creative person take a different direction from the prevailing modes of thought or expression, which is called divergent thinking. It is thought that the frontal lobes are the location where divergent thinking occurs, and the disruption of the frontal lobes with aging, etc. will greatly impact one's ability to proceed with previous acquired abilities in divergent thinking.

The brain is composed of numerous regions that are cytoarchitectonically, cell structure, and functionally distinct. Each is composed of neurons that communicate with one another through axons and dendrites, and it is clear that the types of computations in which individual regions can participate are determined, to a significant extent, by the particular way in which these neurons are wired together (Sporns, Tononi, and Edelman 2000 and Kiehl, 2006). Structural foundations of cognitive and behavioral domains take the form of partially overlapping, large-scale networks organized around reciprocally interconnected cortical epicenters (Mesulam 2000).

The importance of neuronal network ordering or consistent topography arrangements is clear. Universally quantified one-to-one mappings – UQOTOMs - (quantified neuronal cellular anatomical connection framework displaying neuronal inputs and outputs) are computations that connect every unique input of a task with a unique corresponding output important in the basic understanding of the neuroanatomical framework responsible for brain processing. Convergent, divergent, and reciprocal neural connectivity patterns exist to support a distinct sort of computation. In conclusion, investigators propose that topography might, in addition to its role in sensory systems, provide the necessary workable machinery to perform UQOTOMs, and, as such, have a pivotal and foundational role in cognition.

Thivierge and Marcus (2007) have expanded on the importance of cellular topography organization of the brain between the two critical brain regions responsible for thinking. They propose that a key function of cellular topography might be to provide computational underpinnings for precise one-to-one correspondences between abstract cognitive representations. This provides a novel concept as to how the brain approaches difficult problems, such as reasoning and analogy making, and suggests that broadly understanding topographic brain maps could be pivotal in fostering strong links among genetics, neurophysiology, and cognition. This concept, according to these authors, might provide rare and powerful insights into the influence of genes on anatomical formation, neurophysiology, computation, and cognition (Sporns 2000). This expands on this concept by suggesting that functional connectivity may underlie specific perceptual and cognitive states and involve the integration of information across specialized areas of the brain.

Cognitive abilities, such as memory, attention, flexibility, prioritizing, multiple directional simultaneous thinking patterns, and the sheer velocity of thinking are all ascribed to the prefrontal cortex. Clinical studies utilizing PET and fMRI neuroimagry confirm the prefrontal cortex anatomical location as the primary locus for accelerated creative thinking and memory (Fink et al. 2007).

The central nervous system requires the proper formation of precise circuits to function correctly. These neuronal circuits are assembled during development by synaptic connections, or gaps, forming and functioning as bridges between one neuron and another from a dendrite, extension of the neuron, and another neuronal cell body. The process is called synaptogenesis (McAllister 2007). Synaptogenesis and the development of glutamatergic (glutamine is a neurotransmitter) synapses specifically are central to the cognitive process (McAllister 2007; Pennisi 2006). From a molecular perspective, synaptogenesis can be defined as the assembly of hundreds of pre- and postsynaptic proteins into the highly specific structure that form the synapse (McAllister 2007). For glutamatergic synapses to form and function properly, these major components of the synapse—SVs (synaptic vesicles), glutamate receptors, active zone proteins, PSD scaffolding proteins, and trans-synaptic adhesion molecules—must each accumulate at sites of physical contact between axons and dendrites with precise timing.

Given the importance of synaptogenesis for proper functioning of the nervous system, it is surprising how little we know about the cellular and molecular mechanisms of synapse formation. One of the primary reasons we do not understand synapse formation at a molecular level comes from the technical difficulty of studying the formation of individual synapses in intact tissue. Synapses are tiny structures of less than a micron in diameter and are packed into the CNS at an incredibly high density. Estimates range from an average of 200 million synapses per mm^3 in a newborn rat cortex to 4 billion synapses per mm^3 in a five-week-old rat cortex, making it almost impossible to study the formation of individual synapses in intact tissue (McAllister 2007).

The way in which memories are stored or represented in the brain is still unclear. Though much evidence points to changes in synaptic morphology, protein synthesis, and gene expression, the way in which these alterations sustain lifelong engrams remains to be clarified (Mesulam 2000).

These neuronal circuits are assembled during development by the formation of synaptic connections among thousands of differentiating neurons. Proper synapse formation during brain development provides the substrate for cognition, whereas improper formation or function of these synapses leads to neurodevelopmental disorders. Specific neuroanatomical features have been correlated to musical genius (Popp 2004).

The importance of the formation of this unique anatomical substrate from the time of conception forward cannot be overemphasized. This then becomes a unique foundation for cerebration and is clearly unique in terms of the myriad of potential configuration potentials that exist just from a purely structural or "hardwired" standpoint.

NEUROPHYSIOLOGY AND COGNITION

With the formation of the brain anatomical structure, cerebration then is dependent on the communication among the neuronal cellular elements which, as mentioned, are constructed in a very complex and unique fashion. This intercommunication is accomplished by one neuron influencing another or many others simultaneously through a combination of primary electrical events or when one cell signal to another is modified or amplified by specific neurotransmitters important for delivering the message across the synapse, or gap, between cellular connections. Thus, neurophysiologic considerations address this biological process of intracellular communication, which is the basis of functionality and certainly is the biological basis for cognition. The neurophysiologic processes that function on the unique anatomical framework may be more important than the framework itself in terms of determining the functional outcome, especially as it relates to thinking and memory, two functions critical to creativity.

Changes in processing speed may be central to understanding ECA. The neurophysiology involved may be a more important determinant as to functional status than the anatomical framework on which it works. Capacity (anatomical) and processing speed (neurochemical and neurophysiological cerebral mechanisms) are the two major variables important to the cognitive process. Capacity and speed of processing must be unique from day one for the creative individuals.

Multiple observations now lead to the emerging view that changes in neural signaling rather than structural alternatives account for the age-related cognitive deficits, thus it is more of a physiological degenerative process than an anatomic degenerative process (Gazzaley and D'Esposito 2007). These physiological variations combined with the phenomena of neuronal network neuroplasticity, may be the key determinants that separate normal cognitive ability and unique ECA. They may also be responsible for clinical variations of "normal" thinking ability. If this is true, processing speed may be a core attribute at the other end of the spectrum, when accelerated cognitive ability is so apparent, and may be a key differential physiological feature responsible for the difference between average cerebral activities and markedly enhanced cerebral functioning. Perhaps the basic neurobiological framework responsible for accelerated cognition is dependent not only on key structural reciprocal neuronal circuitry that develops over time (Mesulam 2000) but also in large part on the underlying processing speed within these critical circuits. A physiological process, known as top-down modulation, with neural activity enhancement associated with relevant information and irrelevant information suppression thus establish a foundation for both attention and memory processes. This may be greatly enhanced, allowing great focus for individuals with accelerated cognition. Loss of this ability may be a reason for cognitive changes associated with aging (Gazzaley and D'Esposito 2007). Of interest is that the loss of this modulation process may account for aging cognitive changes beyond the structural changes so apparent from pathological examinations.

Changes in processing speeds are a common feature in patients with cognitive aging. Multiple observations now lead to the emerging view that changes in neural signaling, rather than structural alternatives, account for the age-related cognitive deficits, thus it is more of a physiological process than an anatomical process (Gazzaley and D'Esposito 2007).

CHEMISTRY AND COGNITION

There are a few neurochemical considerations important for understanding extraordinary cognitive functioning origins beyond the underlying specific neurotransmitters at the synapse (McAllister 2007) in terms of the types and their concentrations important for neuronal transmission. The neurotransmitter dopamine has been identified as very important in the frontal and temporal lobes in idea generation and creative drive (Flaherty 2005). Mutually inhibitory cortico-cortical interactions via dopamine inhibitory circuits in addition to the mesolimbic dopamine circuits previously mentioned mediate the appropriate balance between frontal and temporal lobe activity in cognition.

Intracellular communication among the neuronal elements of the brain is also dependent on the chemical profile available to assist with the communication between cells. This is accomplished by unique electrical signaling as mentioned above, combined with and dependent on various chemical neurotransmitters to assist and direct the direction of the signal generated in the neuronal cell body. The neurotransmitters make up the message for the signal at the synapse or gap between one neuronal process and another. It is interesting that these same neurotransmitters or neuromodulators that are important for synaptic functioning may be modified or greatly influenced by the environment. Despite agreement on its role in human achievement, the neurobiological underpinnings of creativity are poorly understood. However, myriad cognitive skills exist that are necessary to produce something both "novel and useful," and these skills are manifested within different domains.

Levels of N-acetyl-aspartate (NAA) have been shown to correlate with cognitive ability. MR imaging and divergent thinking measures were obtained in a cohort of fifty-six healthy controls and confirmed a high correlation between NAA levels and degree of cognitive attainment. This is the first report assessing the relationship between brain chemistry and creative cognition, as measured with divergent thinking, in a cohort comparing exclusively of normal, healthy participants (Jung 2009).

NEUROPLASTICIEY AND COGNITIVE FUNTIONING

The neonatal brain is not static and will continue to modify, in terms of the anatomical template and resulting physiological and chemical processes. This may be significantly secondary to a phenomenon known as neuroplasticity (brain changes with time) that may continue for years after birth. Neuroplasticity may occur at multiple neurobiological levels but certainly at the synapse level, both in terms of anatomical hard wiring configuration and the type of neurotransmitters and their respective concentrations responsible for information storage, ease of accessibility, unique reciprocal circuits, synaptic velocity, and neuronal conduction parameters that are important to the end cognitive process. With the accepted understanding that neuroplasticity may have remarkable impact on both hard and soft wiring of the brain with specific influences on synaptic composition, neurochemical makeup, and multiple physiological processes, the role of the environment in cerebral development becomes even more significant.

As Mercado suggests (2008), some species and individuals are able to learn cognitive skills more flexibly than others can. Learning experiences and cortical function contribute to such differences, but the phenomenon of neuroplasticity may be a key differentiating determinant, accounting for wide variations in intellectual abilities. He, Mercado (2008), presents an "integrative framework suggesting that variability in cognitive plasticity reflects neural constraints on the precision and extent of an organism's stimulus representations." The hypothesis is that cognitive plasticity depends on the number and diversity of cortical modules that an organism has available and the brain's capacity to flexibly reconfigure and customize networks of these modules. Both the specific pattern of "hard and soft neuroanatomical wiring" with infinite variations as modified through the process of plasticity and use with age determine in large part the cognitive functioning level at any given time.

CLINICAL IMAGING AND COGNITION

Recent advances in brain imaging has provided a number of valuable clinical tools that give basic neurobiological data in the clinical setting, markedly enhancing our understanding as to how the human brain works. We now can not only identify pathological changes in the brain that help clinically by providing more timely and much more accurate data sooner, but also are able to monitor multiple brain functions in the clinical setting that markedly enhance our ability to understand how the brain operates from a basic science standpoint in real time. Basic brain biology can thus be studied operationally with these image techniques, which vastly expand our understanding of cognition in terms of where, anatomically, in the brain this occurs and delineates critical neurochemical and neurophysiologic aspects of the higher cognitive process.

This is the importance of including a clinical imaging information discussion in the basic science section. Clinical imaging studies, including PET imagery and fMRI studies, can be correlated with clinical studies of thought and memory, which then helps identify the brain regions crucial for cognition (Fink et al. 2007). FMRI studies in schizophrenic populations with auditory hallucinations resembling the experience of having a burst of creative ideas have identified temporal lobes activation, suggesting its role specifically in accelerated creative thinking.

There have been some reports on electroencephalogram (EEG) studies in persons with accelerated cognitive ability, suggesting some specific cerebral neurophysiological changes during the cognitive process (Fink et al. 2008).

EEG studies, in combination with fMRI imagery, in persons with markedly accelerated cognitive ability have identified the frontal and parietal lobe alpha hypersynchronization of both low- and high- band alpha, suggesting active inhibition with creative thinking, left cerebral hemisphere greater than right. (Fink et al. 2007).

Clinical imaging techniques of cerebral blood flow captured during verbal tasks from the Torrance Tests of Creative Thinking have identified the right precentral gyrus, right culmen, left and right middle frontal gyrus, right frontal rectal gyrus, left frontal orbital gyrus, left inferior gyrus, and cerebellum as key anatomical regions involved in the process. This confirms that anatomical regions from both sides of the brain from multiple, distant areas are important and crucial for accelerated cognition in very creative people (Chavez-Eakle et al. 2007). These findings suggest an integration of perceptual, volitional, cognitive, and emotional processes in creativity (Chavez-Eakle et al. 2007).

3. DISCUSSION

Understanding the basic science of human cognition requires a thorough review of all the potential neurobiological variables discussed in the data section. This includes the crucial genetic variables in play along with basic biological processes of anatomical formation, physiological processing, and specific neurochemical composition. All these processes are involved in determining the level of cognition and how it is utilized in the human brain. It is difficult to know which of these multiple variables discussed in the data section are the most important in providing the framework for extraordinary cognitive ability, but it is most likely dependent on a delicate balance of all these components.

The basic issue of anatomical, physiological, and chemical makeup designed by the genome and what part of its design is responsible for exaggerated cerebral energy that appears so crucial for the genius phenomenon to become clinically manifest remains. This most likely is a combination of basic wiring design modified by specific transmitter presence, perhaps in abundance, resulting in much higher performance, perhaps in terms of conduction velocity of thought processes. Is there any hard evidence for this from basic science studies of neurobiological template formation?

The neurobiological template makeup is unique for each individual and is designed by DNA/RNA messaging through genetic inheritance mechanisms. This unique genetic alignment, through the process of emergenesis, which is rarely achieved, is then responsible for the remarkably accelerated neural processing along with the associated heightened inner tension that makes up the core foundation for the highest of cognitive functioning. This framework and associated companion, intense inner tension, are present before birth. The individual must deal with this intense inner tension. It is modified in part by neural modifiers, changing or evolving basic neurobiological template, and crucial environmental exposure, which may have both a direct impact on neural functioning and important indirect influences.

The basic science review must also be focused in attempt to find clear neurobiological template evidence for variations in hard wiring and synaptic configuration that would allow for the suggested marked variability at birth and the rare alignment necessary for extraordinary cognitive functioning.

The basic science of higher cognitive functioning certainly entails a complex interaction of variable and perhaps adaptive synaptic networks as the basic framework is further complicated by variable physiological parameters of timing, direction, intensity, and patterns of neuronal conduction. The transmodal cortical regions, specifically the limbic areas, must provide structural plasticity. This plasticity crucial to the marked differences in basic cognitive potential at birth that is obvious in humans. How much can the basic anatomical framework change with use or lack thereof and lead to a major difference in brain cognition associated with aging? How does this occur and how much precedes birth? Is it a function of adaptation after birth? The combinations and permutations of various structural synaptic variables, combined with secondary variable physiological processing, must be responsible for the great differences we see in cognitive ability from day one in humans. Perhaps the use of these transmodal areas or the lack thereof makes the difference between average intelligence and genius.

What is responsible for the difference, which I believe is present at birth? The genetic influence of these critical, adaptive changes early in development must be very significant. The concept of emergenesis of DNA confluence with primary impact on anatomical and secondary physiological changes must be central to the genius phenomena. There must be a key process that we do not understand very well that links basic DNA (genetic) influence on synaptic anatomical adaption and configuration of physiological network connections. The basic core substance, therefore, is present at birth, with a highly developed anatomical and secondary physiological process in place that is responsible for the accelerated cognitive ability, which involves thought neuronal conduction velocity, ease of multiple network access, storage magnitude, etc. Persons with higher cognitive ability must deal with this heightened basic framework and associated exaggerated inner tension from day one, which is problematic and an opportunity, depending on the surrounding influences available to help them deal with this phenomenon. A clear analogy to computer science is likely, with processing speed reflecting the easy network connections and basic access to storage that may be very large, again based on underlying developmental changes designed by the DNA pattern in the genome.

The human brain, as opposed to more primitive species, has six synaptic levels, allowing an extraordinary potential for a vast spatial expansion of the neural network that links sensation and cognition. Each synaptic level of organization provides a nexus for the convergence of afferents and divergence of efferent influences. This results in the emergence of a large number of alternative trajectories as sensation becomes transformed into cognition. The opportunity for variable trajectories is immense, given this anatomical model, and is central to what separates man from the more primitive species. Memory, at least at the short-term level, is recorded in concert with the visceral or emotional zones of the brain in the hippo-campo-entorhinal complex, but as the memory traces become more long term, they are consolidated through extensive associative linkages. Once memories are consolidated, the binding of their constituents become more dependent on transmodal areas outside of the limbic system, and they become less vulnerable to limbic lesions.

The basic issue as it relates to variable cognitive ability from a basic ana-tomical and physiological point of view relates to the complexity of network available in the human brain. But I believe it also involves a differential ability to use this complex anatomy in terms of processing, such as veloc-ity of transmission, ease of synaptic transmission mechanisms, and basic drive to utilize the network with different degrees of energy. These may be inherited traits. One's basic drive, energy of use, synaptic ease, neuronal velocity, etc. may be inherited through the DNA, and thus one's ability to take advantage of the complex anatomy available may be vastly different from one individual to another. Do we have evidence for different neuronal velocities from one individual to another? I do not think so, but it may be postulated as a factor, along with the basic drive and motivation to utilize the anatomy in an ever-increasing amount creatively. I am not sure we will ever be able to clearly show differences in these basic use parameters in cognitively advantaged persons, but it is clear that at the time of birth, these persons have a different cerebral energy that separates them from others. The basic theory is therefore that cognitively advantaged persons have inherited a clearly advantaged "use" mechanism for cerebral cognition that may reach astounding heights, given the available complex anatomy framework that can be utilized.

The DNA passage, thus the genetic influence, is paramount to any discussion as to the origins of genius. How is this expressed in a basic anatomical format? It is doubtful that the geometry of the cortical structures themselves is the key marker for various levels of inquisitiveness, so the difference must be reflected in the physiology or function of the neuropil.

This is to say that the DNA influence as it relates to higher cognitive function is configured in the physiology rather than in the anatomical structure of the brain, although this has been argued. If it is the physiology, is it the velocity of neurotransmission, the velocity across individual synapses, the number of the synapses in key areas, or the amplitude of the signal. Or does the physiology over time change the circuitry that is not recognizable with anatomical study but the pathways are truly different allowing for more rapid and intense thinking and allowing for extraordinary memory? Genetics clearly play a pivotal role in determining brain function as it relates to higher cognitive functioning. This is evidenced by clear indications of a different cerebral energy source very early in life that goes well beyond any environmental influences that could be in play at that early age of development. From biographical observations, there seem to be a high cerebral energy and constant that very bright individuals have to deal with from day one. They are clearly different from their peers. This then becomes a tremendous asset if properly channeled by environmental influences, but even when properly channeled, it carries with it a heavy burden for the individual to engage and relate in a "normal" way with individuals who surround them.

If healthy environmental influences are not present and unfortunate negative modifiers are in place, the individual with ECA potential may be burdened with the lack of intense focus necessary to relieve the tremendous energy they feel. This may, in some instances, lead to very important negative consequences for the individual both in terms of the degree of intellectual focus and what form the focus takes.

New synthesis theory involves a concept of the individual human brain constructed of complex neural circuits that start taking shape before birth and continue to grow and change throughout life. This is the phenomenon of neuroplasticity. Gene design and cell maturation are therefore influenced by the environment, experience, and cultural exposures in a dynamic fashion. The brain, therefore, is not completely hardwired and through the process of neuroplasticity can be significantly molded over time in many different directions, depending on the exposure to significant modifiers both neurobiological and environmental in origin. The neurobiological framework is uniquely designed for each individual via genetic DNA and RNA messaging and conveys the basic design as genetically determined by the configuration of the genome. Extraordinary cognitive ability results from a unique configuration of the genome; extreme tension results clinically, perhaps secondary to high-velocity neuronal transmission velocities along with other variations in the basic anatomical and physiological makeup, including critical variations in synaptic configurations that were discussed in the basic data section. The basic configuration involves more than just increased nerve conduction times but also includes unique, intricate anatomical framework changes with very complex neuronal network patterns. The biology may be significantly different from "normal" at conception, which, combined with critical early developmental environmental exposures, may set the stage for high levels of tension that always seems to be a partner with extraordinary cognitive ability. Inheritance therefore becomes crucial for two reasons. The first reason—and most important—is the basic brain framework design or neurobiological template. The second reason is the presence or absence of a robust, specific area of interest, literature, art, music, etc, in past generations that may then be a major environmental factor influencing the individual as to the best direction for his or her intense intellectual energy.

4. SUMMARY

The combination of complex synaptic anatomical pathways primarily in transmodal cortical areas with the variable physiological networks that develop in this framework, both of which can be unique per individual, is responsible from a basic science perspective for higher cognitive ability. This, in my opinion, occurs before birth and may well be secondary to the DNA or genetic influence on brain anatomy, chemistry, and physiology. This development in the unique setting of extraordinary cognitive ability may be the result of a genetic mechanism known as emergenesis. Emergenesis is a unique and likely purely random combination of genes that may lead to qualitative shifts in capacity and intellectual ability. The phenomenon of cerebral plasticity of the anatomical substrate and physiological networks may have an important role as well, even in the prenatal setting of brain development. In certain individuals, all these components come together at birth in a very unique fashion, I believe, providing the basic template for extraordinary cognitive functioning. The resulting heightened cognitive ability in these persons is therefore very sensitive to this basic framework at birth. The real challenge is what happens next in terms of the individual's ability to cope with this heightened intellectual energy, all of which is very sensitive to the surrounding environment and the ability of the individual to handle the emotional hardships always present with this phenomenon.

The basic neurobiological template at birth represents a unique development designed by the DNA makeup of gene combinations at conception that are responsible for the structural or anatomical architecture of the brain and the physiological and chemical processing critical for cerebral functioning. In trying to better understand accelerated cognitive ability, numerous basic science observations have been made that may give some insight to its origins. These include critical genetic data, along with important microscopic and macroscopic anatomical information, combined with physiological and chemical processing differences that may help explain the unique variations responsible for the unique biological template at birth responsible for extraordinary cognitive functioning.

Some of these observations may be more important than others are, and it is clear that we still do not understand the key basic elements critical for accelerated cerebration. It is clear, however, that the basic neurobiological template or framework for accelerated cognitive functioning is, by and large present, at birth, with the exception of some plasticity changes occurring after birth and well beyond. All of this is to be significantly affected by environmental modifiers that will be discussed in the following sections.

PART IV:
HISTORICAL BIOGRAPHIES AND CLINICAL SCIENCE OF ECA

A: BIOGRAPHIES

INTRODUCTION

This phenomenon of markedly accelerated cognitive ability is more clearly understood and appreciated with detailed individual biographies as specific examples of accelerated cognitive behavior. These individual studies will be coupled with comparative observations as to areas of commonality and areas of dissimilarities that are important to better understanding this phenomenon in man. Obviously, the outcome of this cognitive variation is in individual behavior and perhaps unique performance, which may significantly impact contemporary society and multiple future generations, given the potential power of the basic accelerated cognitive ability.

The basic underlying neurobiological template responsible for the accelerated cognitive ability may be quite similar for all these individuals. However, the factor or factors responsible for the specific focus or direction of the intellect that mostly develops very early in life becomes crucial in determining one of three potential outcomes. The three directions are positive societal influence, negative societal impact, and potential primary significant psychiatric disease in the absence of either a direction or focus, all of which have serious individual and societal implications. Primary biographical features to be emphasized with these biographies include:

- Lineage
- Birth location and immediate environs
- Primary nurturing features
- Age of awareness and under what circumstances
- Timing and stimulant for primary focus
- Psychological issues, both early and late
- Degrees of comfort versus discomfort
- Lifestyle and work ethic
- Environmental exposures
- Physical and mental health
- Insights via education or nurturing in general
- Legacy of the specific focus chosen.

These individual legacies to present and future generations may either be quite positive with gifts of science, art, and leadership etc., which are so valuable to understand and enjoy or may in some instances lead to quite negative or destructive legacies, again depending on the ultimate focus of the individual intellect.

It is likely there is a third, less understood group that may be even of more practical importance given the numbers involved: individuals with this same cerebral gift of accelerated cognition, who, for various reasons, never move in either a positive or negative direction and find themselves unable to focus or direct their intellectual energies. There are many reasons for this, not the least of which is the universal significant inner tension that all these individuals must deal with.

These biographies have been chosen to highlight a variety of lives but are written so they can be compared for commonalities and dissimilarities of the many important factors in the underlying biological process and the subsequent behavior thought to be strong determinants of outcome. These biographical discussions are not intended to be complete, detailed biographies. They focus on the basic intellectual gift, potential origins, important early environmental influences, behavior as a result of this gift, the emotional burden that comes with it, interaction with peers, methodologies devised to deal with these burdens, health implications, lifestyle, and finally a brief comment as to the legacy or the imprint that the individual has left for future generations.

The underling principal thesis is the basic neurobiological variation accounting for greatly accelerated cognitive functioning has the same basic biological elements for all these individuals, and the difference are played out in large part with variable very early environmental exposures or modifiers that determine the direction or the focus and its intensity. Thus, attention to lineage, family composition, and close peers very early, as well as the developmental years, emotional observations, interpersonal relationships, work ethic, lifestyle, health issues, peer input, and, of course, legacy of individual efforts all are important as specific areas of comparison.

There appears to be a great deal of commonality between these individuals, especially very early on, which is evident with closer examination of the biographic data. This data may be very important to better understand the roles of inheritance and environment in combination, which is crucial to the ultimate outcome of this unique cognitive gift.

One clear and important clinical feature is in common: an intense degree of inner tension coupled with a unique curiosity from day one. This is an integral part of the phenomena of accelerated higher cognitive functioning. These studies, therefore, should serve as sample clinical outcome studies of the underlying neurobiological phenomena under discussion to be better understand the nature of the process, in combination with the basic science and clinical studies detailed in this study. A few biographies are mentioned in a later section. These individuals perhaps have similar intellectual attributes that utilized their enhanced intellectual energies in directions that were destructive to society. These discussions will also include brief comments as to the potential third group, who are less well understood but have the same underlying biological mechanisms in place yet do not launch their intellectual energies in either a positive or a negative direction with great focus. They still have to deal with the severe inner tension that is part of the underlying neurobiological process responsible for accelerated cognitive ability. This potential outcome will be addressed in a separate section.

The potential implications of this will be discussed as to issues of the importance of early recognition, environmental and educational exposures of importance, psychopathological implications, and behavioral patterns of these individuals of general societal importance.

JOHN DOE BIOGRAPHY AS MODEL LEAD INTO THE BIOGRAPHY SECTION

The John Doe modern fictional biography discussed in parts 2 and 3 serves as an overall introduction to the ECA phenomenon, along with specific illustrations of some of the important basic scientific issues deemed central to this discussion as it relates to better understanding its possible origins. The biography was constructed to emphasize the common features individuals with ECA seem to have. A review of this biography, therefore, will greatly assist us as we now examine individual biographies from multiple genres with an emphasis on many of these common features and key dissimilarities.

The focus of the following biography presentations will be on their unique cognitive abilities and the features in common. John Doe represents a fictional and more modern biography, serving as a lead-in to the multiple biographies discussed below as a way to further understand the phenomenon from a real-life standpoint.

"I was cut off from the world. There was no one to confuse or torment me, and I was forced to become original."
Franz Joseph Haydn

FRANZ JOSEPH HAYDN

INTRODUCTION

Franz Joseph Haydn, 1732–1809, was a prolific composer of music. His works included 118 symphonies, 115 Masses, and two wonderful oratorios completed near the end of his life. He was not only recognized for his remarkable musical creative ability during his life time but also continues to be revered for his massive legacy of musical repertoire that continues to be loved today whenever and wherever classical music is appreciated.

He clearly was a musical genius from a very early age. He had an astonishing ability to write wonderful masterpieces and continued to energetically produce a remarkable volume of work year after year, seemingly without end. He lived a long, relatively secluded life in a demanding environment and was generally calm with a level disposition. This offers some interesting contrasting points of comparison with other persons of great cognitive ability.

BIOGRAPHY

Lying on his deathbed, looking at a picture of Haydn's birthplace, Beethoven remarked, "How it was possible that so great a man could spring from this barn?" (Jacob 1950). The barn was located in the village of Rohrau, on the border of lower Austria and Hungry. It was 1732. His father and forefathers were all artisans, primarily Cartwrights, living a very simple rural life in an agricultural setting close to nature. His early life was one of a simple peasant life combined, perhaps most importantly, with a very early, unusually keen awareness and curiosity for nature.

Both of these items set the foundation for who Haydn became and provided him with all the important sensitivities so central to his latter creative compositions. His birth date was either March 31 or April 1, 1732, and he was baptized Franz Joseph Haydn. As a child, Haydn developed an eternal alliance with nature. He grasped the acoustic language of nature at a very early age, and the sensory experiences of his childhood became part of his very being (Jacob 1950).

He was born a German, was Austrian by nationality, and had some Croatian background. Thus, a melting pot of races and influences perhaps accounted for his broad musical background, which he brought to life in his compositions.

Extensive genealogical research has not identified any individuals on either side of his family tree who were musicians or particularly intellectually inclined. He did, however, have a deep religious sense, tenacity of purpose, and a great deal of passionate desire to improve himself and his family in their station in life. All worked with their hands. Fortunately, Matthias, his father, was musical and played the harp in the home. Haydn learned and sang all these songs by age five. He apparently had perfect pitch and a beautiful voice in addition to learning how to play the violin at a very early age. Very little is known about his mother, Maria, other than she wanted him to become a priest. She bore twelve children, six of whom died early in life. Haydn was the oldest surviving child of the family. Like Mozart, he never forgot a note because he had an astonishing memory, but, in contrast with Mozart, he was not precociously creative and did not mature into creating musical compositions until much later in life, in his early twenties. Johann Matthias Franck, the husband of Mathias Haydn's stepsister who lived in Hainburg, recognized the young Haydn's ("Little Sepperl") musical talent and suggested that he come to Hainburg to get a proper education. He did, at age five and a half, never to return home. He apparently never had a formal education outside of what he was later exposed to at St. Stephen's Cathedral in Vienna and was never exposed to worldly knowledge. He did not learn to write in the German language.

In 1740, at age eight, he was recruited as a choirboy for St. Stephen's Cathedral in Vienna. He continued to do that for the next nine years of his life. This early exposure to music may have provided him with that vital environmental stimulus that evolved into his focus of musical composition. His voice changed, and he was released from St. Stephen's at age seventeen. He was replaced in the choir by his younger brother, Michael Haydn. These years at St. Stephen's were a very austere existence for him, but he was taught reading, writing, arithmetic, religion, and Latin in addition to singing. He had exposure to the violin, with a great deal of focus on piano lessons. He became skilled in the piano and the violin and learned a lot about musical composition. He lived at the old St Michael's house near St. Stephen's Cathedral. He devoted his time to understanding musical texts and decided that he wanted to be a composer. He learned Italian, the world's musical language at the time, which accelerated the development of his musical sophistication.

He was not a fast learner, but he was self-taught as reflected in studying the original textbooks and using margin notes to reinforce his learning. He studied Carl Philipp Emanuel Bach's book of 1753, *The Essay on the True*

Manner of Playing the Clavier. Gluck advised him to go to Italy to complete his musical education, but Haydn did not do so.

His age of awareness of his special gift came relatively late in is life, after his time in Vienna at St. Stephens. His focus remained on music from very early on, with the influence of his father's interest in music and an early musical tutor. Haydn had a very even temperament during his entire life and was generally calm, which may have been an important factor leading to his long life of seventy-seven years. There is no evidence of wide mood swings, and he seemed to relate well with peers, employers, and society in general. There was no sense of psychological discomfort; he knew very well who he was and what he was capable of, and he had considerable confidence in his composition skills.

He enjoyed a great deal of accolades for his work from not only his peers but also the musical audiences of his day. This specific profile represents a marked deviation from others with extraordinary cognitive ability who had generally painful, short lives without positive social interaction and who were consumed by considerable inner tension.

At age twenty-five, Haydn wrote his first string quartet, and at age twenty-eight, he wrote his first symphony. He was hired by a great Bohemian aristocrat, Count Maximilian Morzin, as a musical director at a salary of 200 gulden with free room and board, including wine. He worked eighteen hours per day with compositional work at night. Haydn was then hired by the Esterhazy estate, Paul Anton Esterhazy, the reigning prince, after Count Maximilian Morzin developed financial problems. Haydn's title was to be Vice-Kapellmeister with, according to point four of a fourteen-point contract he made with the count, the obligation to compose such music as his Serene Highness may command and neither to communicate such compositions to any other person nor to allow them to be copied, but to retain them for the absolute use of His Highness, and not to compose anything for any other person without the knowledge and permission of His Highness (Jacob 1950). This contract was consummated on May 1, 1761.

His position with the Esterhazy estate gave him the opportunity to focus on music rather than worry about making a living and thus allowed him time to be creative. The great demand from the prince himself resulted in a high volume of compositions during the years. He worked very hard during his life, part of which was secondary to the constant demands of his employer to compose music for him—a fact made clear in the original contract language. He was very engaged as a teacher, conductor, and composer and must have used his time very efficiently to achieve all that he did.

The innovator of the modern symphony was Johann Stamitz of Mannheim, Germany, and Haydn took up where Stamitz left off with his early death in 1756 before the age of forty. Haydn moved the symphony from the Sinfonia form to the full symphony and moved the genre out of the theater. Haydn's teacher was C. P. E. Bach, and he soon became the father of modern orchestration. Most of the symphonies that he wrote between 1770 and 1780 present moods of melancholy succeeded by strong passion. When Haydn and his wife, Anna Aloysia, came to Eisenstaedt in 1761, they lived, as did the other members of the prince's band, in the "music building," a stately edifice near the church. His marriage with Aloysia deteriorated over the years to a point that his life as a bachelor after her death became much more attractive to him.

Haydn was the prime originator of the string quartet and greatly expanded the genre. The string quartet was like writing a diary for Haydn, into which he put his moods and experiences from the moment. Haydn was a genius of tranquility and thus not a good candidate as a composer of operas.

Haydn clearly worked very hard and was very responsive to his employer's needs in terms of new unique musical compositions, resulting in a vast number of musical works. He spent long hours composing, and he directed the Esterhazy estate orchestra. He did not have a happy marriage, as mentioned, and apparently was not understood or honored by his wife. His environment was quite unique for an artistic "genius," with a protective secluded existence for many years that allowed him time to do his work without monetary concern as long as he continued to produce new compositions at the pace that Prince Paul Anton Esterhazy demanded.

Haydn and Mozart had a remarkably close friendship. Haydn was quoted as saying, "Ah, if only I could persuade every friend of music, but especially the great ones, to understand and to feel Mozart's inimitable works as deeply as I do and to study them with as great feeling and musical understanding as I give to them. If I could, how the cities would compete to possess such peerlessness within their walls. Prague would do well to keep a firm grip upon this wonderful man—but also to reward him with treasures. For unless they are rewarded, the life of great geniuses is sorrowful and, alas, affords little encouragement to posterity to strive more nobly; for that reason so many promising sprits succumb...it angers me that this unique man Mozart has not yet been engaged by some imperial or royal court. Forgive me, honored sirs, for digressing, but I like the man too well" (Jacob, 1950).

Mozart and Haydn's friendship was only second to that of Goethe and Schiller. Haydn had a much different temperament than Mozart, as he was in harmony with himself and reminded one more of a painter than of a musician. He was kind and gentle, was not highly emotional, and was destined to live a long time. He was not precocious as Mozart was, and was slow to mature. Haydn was always audiologically sensitive to the external world, especially the sounds of nature. Mozart, on the other hand, was more intoned to his inner emotions.

The abolition of serfdom struck a mortal blow at the wealth of the Esterhazys and other landed aristocrats. Nicholas died on September 28, 1790, and thus Haydn was free after thirty years. He was thinking of going to London, but it was hard for him to adapt to that concept after thirty years of living secluded at the Esterhazy estate. He did go to London in late December 1790, after saying good-bye to Mozart. He was very modest and timid, which took his London hosts by surprise, as they thought he would be much more outgoing and forceful. He was not proud and arrogant as most geniuses are. He was sixty years old at the time, and his music was an instant success in London.

Haydn became a disciple of Mozart, and after Mozart's death (December 5, 1791), his London Symphony of 1792 no. 98 contained an adagio cantabile in the second movement that many think is a requiem to Mozart. His relationship with Beethoven was very strained, but they did respect each other, although Beethoven realized early on that he was not learning anything from Haydn, but he still kept seeking instruction. Haydn's last six symphonies were quite serious and melancholy in nature. The mood was one of parting from a century of security. The London symphonies were his last. After them, he turned to oratorios, such as *The Creation* and *The Seasons.*

Instead of settling down in London, he returned to the Esterhazy orchestra and Nicholas II, the fourth prince whom he had served at Esterhazy. He felt very competitive with Handel and took Lindley's poem relating to the creation and created music for it after having learned that Handel turned down the urging to create music for the poem. By that time, Haydn went back to work for Nicholas II and had a great deal of innate self-assurance. He was able to have a sense of independence and self worth. Reflecting this, Haydn's chief works during this time were *The Creation* and *The Seasons,* which were not written on order from Nicholas II but composed by Haydn as a "free enterprise." He was only obligated to write one Mass a year during this time, and it took him only three months to do so. The first performance of *The Creation* was on April 30, 1798, in the palace of Prince von Schwarzenberg. He stated that he had not created the masterpiece but a power above was responsible for it.

Haydn had generally good health all his life with an orderly daily regimen of work, abstinence from all excess, and a sturdy heritage from his ancestors. He did develop headaches for the full two years that he created *The Seasons* and was exhausted at its completion. He gave up his post as Kapellmeister in Eisenstaedt in 1803 because of his age of seventy-one and found that he could not compose any longer. He attributed his difficulties to a "nervous condition." He continued to have many new ideas but could not translate them into musical compositions. He did catalogue his compositions, which included 118 symphonies, 83 string quartets, 19 operas, 5 oratorios, 24 trios, 163 baritone pieces, 44 piano sonatas, 115 masses, 10 smaller ecclesiastical compositions, 24 instrumental concertos, 42 German and Italian songs, 39 canons, 365 Scotch and Welsh songs, and numberless capriccios and divertimenti, etc (Jacob 1950). He was always learning and experimenting with his music. Could he have created all this wonderful music without the security and relative isolation of the Esterhazy estate for thirty years, coupled with the musical demands of his prince?

Haydn died on May 31, 1809, at one o'clock in the morning. Johannes Brahms later rediscovered the greatness of Haydn's works as a composer and orchestrator.

HAYDN AND ACCELERATED COGNITIVE ABILITY

Haydn clearly possessed extraordinary cognitive ability at birth, as revealed in his very early unique curiosity and lifelong need to remain engaged intellectually. He applied remarkable energy to his chosen focus of musical composition and created music independently for many years. He was driven by his need to perform as directed by his employer but coupled with his innate intellectual energy and ability that allowed him to complete so many outstanding compositions. As is typical of those who possess extraordinary cognitive ability, he valued his alone time for creation but, atypically, led a very ordered, calm, and unemotional life and lived many years without major illnesses. This may have occurred because of his rather unique living situation as Kapellmeister at the Esterhazy estate. He clearly had the fortunate combination of a clear focus, demanding environment for production, and the intellectual energy and creativity to successfully accomplish to the extent that he did. Key observations as to insights into the phenomena of extraordinary cognitive functioning gained from his life include:

- Clear early unique curiosity
- Early environmental exposure to music
- Not needing early formal education
- Stimulation as to production and reinforcement for success
- Acceptance of his lifestyle with good management of his time and absence of anxieties with a long, illness-free life
- Lifelong curiosity and desire to continue to compose or expand his chosen musical passion.

HAYDN: FEATURES OF COMMONALITY AND DISSIMILARITY WITH OTHERS WITH ACCELERATED COGNITIVE ABILIY

Characteristically, Haydn was very curious from day one and was extremely sensitive to the sounds and beauty of nature, which provided him with a very early important focus for his intellectual curiosity. Like Chopin, he was fortunate to have a very early exposure to his eventual passion, music, in the home with his father playing the harp. He worked extremely hard all his life and had an obvious innate skill for musical composition. His higher cognitive ability was not an obvious family trait, as is common with these individuals in general. He was remarkably productive throughout his life, given his work ethic and his innate intellectual abilities.

There are two clear areas of dissimilarity from other creative individuals, including a long life without major illness that was associated with a rather calm temperament, social ability, and recognition of his talent by peers and the public alike early on in his career.

HAYDN AND HIS LEGACY

The vast musical repertoire he created during his long life remains his legacy to all future generations, and his importance as a mentor to many classical and romantic composers who followed him is obvious and cannot be over emphasized.

Haydn's musical compositional genius is quite clear, and it is important to point out that there was no history of remarkable musicians in his family nor was there any clear aspect of his early life that seemed central to his development other that his keen sensitivity to nature at a very early age and the very important prolonged exposure to the St. Stephen's boy's choir in Vienna from age eight to seventeen. Fortunately, he discovered his passion for musical composition very early in life and applied his enormous energy to creating music in multiple genres. His calm, well-ordered life may have been very important for the long, productive life that he had.

He had no formal education or clear mentors, other than his attraction to the written work of C. P. E. Bach The extreme internal tension created by the ECA was successfully channeled and allowed to flourish. His long productive relatively calm life in a secluded, protective environment is unique when comparing with other creative geniuses, but his very early keen awareness and fortuitous environmental exposure seem consistent with the importance of these features to the successful development of a creative genius.

"Neither a lofty degree of intelligence nor imagination nor both together go to the making of genius. Love, love that is the soul of genius."
Wolfgang Amadeus Mozart

"I pay no attention whatever to anybody's praise or blame. I simply follow my own feelings."
Wolfgang Amadeus Mozart

Wolfgang Amadeus Mozart

Introduction

Wolfgang Amadeus Mozart was born in Austria in the mid-eighteenth century. He had one of the most powerful intellects that the world has even known. Fortunately, he was able to channel all this intellectual energy into creating beautiful music, leaving the world a priceless legacy that remains a cherished gift for the ages. The origin of his intellectual prowess remains a mystery, without any clear genealogical markers with the one exception. His father, Leopold, was a capable violinist and composer. Mozart's gift may well be attributed to a combination of a critical alignment of genetic determinants or the phenomenon of emergenesis discussed in the basic science section, combined with exposure to unique environmental modifiers very early in his life accounting for the flowering of his genius.

As is common for most persons with these unique intellectual gifts, there was a clear and significant disconnection between his extraordinary mature inner artistic creativity and his external interactions with his peers and society in general. Like other persons with intellectual gifts of intense cerebral energy, he certainly enjoyed an amazing gift but also had to deal with severe personal burdens. His life was short and accompanied with considerable discomfort, given the magnitude of the disconnection between his mind and his daily social interactions. His musical creativity and gift remains his powerful legacy for the ages. Most people appreciate and are aware of that gift, but few know or are concerned with Mozart the man and the intensity of the discomforts that he had to endure to achieve his greatness.

BIOGRAPHY

The life of Wolfgang Amadeus Mozart begins on a Sunday morning in Salzburg, Austria, on January 27, 1756. His father, Leopold, was originally from a family of bookbinders living in Augsburg, before he moved to Salzburg to pursue his musical career as a violinist and court composer. Mozart's mother, Anna Maria Bertland, was the daughter of a local court official and did not have any known musical abilities. She, however, is fondly remembered for her cheerful and fun-loving personality, which she gave to her son, enabling him to gracefully carry the "burden of genius" (Davenport 1932).

His only surviving sibling was Nanneral (Marianne), an older sister born in 1751. There are no other members in Mozart's genealogical history with musical aptitudes or high-intellectual pursuits in other areas beyond, of course, that of his father, Leopold. Mozart was born into a home whose major focus was music, with frequent guests either players of instruments or composers in their own rights. Leopold, Nanneral, and Wolfgang all played musical instruments at a very early age, and it is believed that Mozart was able to write musical notes before he could write words. His instruments were the violin, which he started playing at the age of three, and the clavier.

His whole life quickly became music, triggered in part by his father's early recognition of his extraordinary ability to not only read and play music but also compose music at a very early age. Leopold was his first music teacher but stated early on that "it is plain that this is not so much a lesson as a checking-up on what wonder God has wrought since yesterday, as he was already convinced that the 'genius' of his son was the mask of divine favor to be reviewed and guided" (Davenport 1932).

With this awareness, Leopold becomes his promoter and thus began to arrange for performances by both of his gifted children. Wolfgang and Nanneral played for the Elector Maximillian Joseph III at his palace in Munich and for many other noble personages throughout Europe. Leopold continued to seek out the nobility and took his family on several journeys to promote the talents of the young Mozart children, determined to take advantage of the potential wealth and recognition of their prodigious performances.

After the very early trip to Munich, the family visited Vienna, including a performance for Empress Maria Theresia at the Schonbrum. The third trip, when he was seven and Nanneral eleven, was to Paris in 1763, with several stops between Salzburg and Paris, including one at Mannheim where he and Nanneral played for the Elector Palatine Karr Theodore. The elector had the "best" orchestra in Europe and eventually made a very important contact for Leopold with Friedrich Melchior Grimm in Paris, who opened many doors in France, including concert appearances at Versailles.

This trip left a great impression on the young Mozart but was not financially successful for the family. He wrote some of his first works in Paris, including two piano sonatas with violin accompaniment. Later in London, he wrote his first symphony, K 16, at age eight while attending the court of King George the III. A bit later, he wrote three more symphonies interspersed with numerous piano and violin sonatas. He also met Johann Christian Bach while spending one year in England. Leopold then decided that they had spent enough time in England and returned to Salzburg after a three-and-a-half-year absence.

They left England for the Hague in September 1765, and while in Holland, both children became very ill with high temperatures. The next adventure was to Italy, the center of the musical world, when Wolfgang was fourteen. Time was primarily spent in Naples, the center for Italian opera. He was at the house of Sir William Hamilton, the British ambassador, almost daily, playing the clavier remarkably for his age.

"His 'genius' for melody was as innate as his gay disposition, and his magnificent musical thinking a natural gift. His marvelous aptitude for characterization in song appeared later still, after he had created a place of his own in the musical world" (Davenport 1932). He became quite critical at a very early age of other musical performers, which earned him many enemies for his frankness. Mozart was not polite and was very blunt in his social interactions. He was accused of being tactless, impatient, and sarcastic.

He had a very stressful relationship with his father, Leopold, who remained overbearing and extremely critical and never released him the rest of his life. Leopold was also very critical of his romantic attractions for Aloysia Weber in Mannheim and demanded that he move on to Paris where "the great folks were." This was in 1778. His favorite musical instrument at this time was the organ, and he had decided opera was his most favorite genre for composition.

Mozart's mother died while he was in Paris, thus adding to his overall misery with the city, which was Leopold's choice for him to continue his musical career. He did not want to return to Saltsburg after three years in Paris, so he went back to Mannheim and Munich and eventually on to Vienna on March 12, 1781, invited by the Archbishop of Saltsburg, Count Hieronymus Colloredo, to accompany him on his journey there. He remained in Vienna where he met Joseph Haydn, whom he considered the "true master."

At age twenty-five, he moved out of the archbishop's residence and moved to the lodging of Cacilia Weber as a renter. He wanted to marry Constanze, Aloysia, Weber's sister, but his father was fervently opposed and would not give him his permission.

During this time, Mozart would compose for three hours in the morning and then give lessons to the nobility in the afternoon, charging six ducats for twelve lessons. He generated income from musical lessons and concert performances. His major compositional interest remained opera, and he composed *The Abduction from the Seraglio*, a German opera based on Italian opera buffa, and had his opera, *Die Entfuhruing* open on July 16, 1882.

It is well known that Mozart worked incredibly fast, that he worked out everything in his head, down to the last detail, before setting pen to paper, and could then write it all out in one sitting (Braunbehresn 1986). It is also well known that he always needed clear motivation, a commission, or a specific occasion for performance to trigger his creativity. Apparently, only the deadline pressure seemed to release his artistic power.

He married Constanze Weber on August 4, 1782, at St Stephen's cathedral in Vienna without his father's permission, who remained very bitter about this the rest of his life. He thought that the marriage would prove to be the ruin of Mozart's career. Wolfgang's music took a turn for the immortal about the time he married her, which had nothing to do with any possible inspiration furnished by her. In Vienna, he met Antonio Salieri, who became quite competitive with him. He worked with pupils such as Johannes Hummel but could not support himself giving lessons so he continued composing and playing concerts while he awaited a favorable appointment that he was sure would happen soon. His periods of pianoforte performing were thus always marked by voluminous clavier composition, because he wrote new sonatas and concertos for each performance.

He often went to quartet parties and so became the source of much of the great chamber music of the time. He related to Joseph Haydn during his visits to Vienna from the Esterhazy estate and became a close friend and mentor. Mozart ended up dedicating six quartets to him. Constanze and Wolfgang returned to Salzburg for three months to be with Leopold and Nanneral after the loss of their first child, but the visit did not go well with difficult relations between Constanze and Wolfgang's family, so they returned to Vienna and never returned to Saltsburg after that. He did compose the Linz Symphony in C Major (K425) in four days so he could perform at Linz on the way back to Vienna. Then began the brief but miraculous years of 1783–1784, where his musical prowess peaked in Vienna. During 1784, he wrote six piano concertos, one piano quintet (with winds), one string quartet, two sonatas, and two sets of variations for piano plus a few smaller compositions. He joined the freemasonry movement at the time, primarily identifying with the concepts of training oneself, practical humanity, and tolerance. This organization at the time attracted most of the freethinkers and critical minds of the age by virtue of its humanistic philosophy. He was not well received by peer contemporaries, as they did not understand his musical style and feared what they could not understand.

Wolfgang was very busy with concerts and the composition of wonderful piano works. He was expected to write new concertos or set variations for each of his concert appearances, which he did under a great deal of pressure to compose. This continued his pattern for the rest of his life. Haydn was very impressed with Wolfgang's abilities as a composer and told Leopold that "he was the greatest composer he had ever known or only known by reputation." (Davenport, 1932) Salieri made it clear, however, that he did not think that Mozart was capable of composing an opera. However, Wolfgang and Lorenzo Da Ponte began on *The Marriage of Figaro (K492)* in 1786. It was always clear that his first love in music was opera. The characters of Figaro, and later, of Don Giovanni, came alive in music and were perhaps the first characters ever to spring forth from this medium. Mozart and Lorenzo Da Ponte were a combination of composer and librettist and were a combination whose equal has never since been reproduced.

The Marriage of Figaro was performed for the first time in Vienna in front of Joseph and was a tremendous success, but it was perceived as just a passing fancy and was not repeated many times. Wolfgang continued with problems making enough money to comfortably live on as his primary source of income remained piano lessons that he gave. However, he made a career for himself as a freelance composer and virtuoso pianist, supplementing his teaching income. He never did obtain a "court position" which had been his lifelong goal. He liked to live high and dressed as the nobility did. Leopold was asked but declined to take his children when he and Constanze wanted to go to England. He wrote three symphonies—39, 40, and 41—in eight weeks after this bitter disappointment. The dramatic G Minor is supposed to be the culmination of all the tragedy and frustration in his life. The Jupiter (Symphony no. 41) is the salute to the future and promise of the next century. The beautiful E Flat was to represent Mozart's farewell to his youth (Davenport 1932).

The Marriage of Figaro was a great success in Prague. He, as a result, decided to take his family there and wanted to compose another opera and dedicate it to Prague. This was in 1787, when he met the seventeen-year-old Beethoven. He and Lorenzo composed the opera Don Giovanni while in Prague, where he was greatly appreciated. He was thrilled by this unique experience of being fully appreciated for the only first time in his creative life. The music of Don Giovanni was written in the midst of "good company, chatter, punch, games, and dancing." (Davenport, 1932) "To come back from composing the Jupiter, Don Giovanni, and the Marriage of Figaro to a small, homely, nervous man, worrying about his debts in a shabby suburb garden, is to see plainly the working of the infinite."(Davenport, 1932)

He continued to be burdened with a mountain of debts aggravated by expensive cures for his sick wife (Davenport 1932). During the last year of his life, his mind and spirits were not reinforced and refreshed. His poor body withered, and his brain faltered. His health began to fail at about this time, and it is clear that his physical condition was influenced by his financial circumstances. He did write a requiem Mass on a mystery commission, which later became a requiem to his own death. The patron who commissioned the *Requiem Mass* was Count Franz Walsegg-Stuppach, who wanted to use it to celebrate the death of his wife Anna, who had died in February 1791.

At about this time, he began to have fainting spells and headaches plus he became very nervous. He died on December 5, 1791, with massive swelling in his hands and feet and with great pain and immobility thought to be secondary to an attack of rheumatic fever with both cardiac and renal involvement. He may have had nephritis terminally but was treated for meningitis (Braunbehresn 1986). He remained with financial distress to the end of his life, as evidenced by the fact that he only received $200 for *The Marriage of Figaro*, $225 for *Don Giovanni*, the same for *Die Zauberflote*, and about $112 for the *Requiem Mass* on commission.

Mozart and Accelerated Cognitive Ability

Mozart may be the purest example of emergenesis, previously described in the basic science section, without any clear genetic clues in the family as to the origin of his genius. He was acutely aware from a very early age that he had an inherent ability in musical composition and performance. Mozart displayed multiple characteristic features of extraordinary cognitive ability in his life. From a very early age, he was intellectually quick, focused if not driven to create musically, had a keen sense of humor, displayed impatience with peers, and was quite outspoken on most matters.

He remained focused and was able to compose under very difficult circumstances, producing large amounts of compositional work when most innovative persons would be unable to create given the "background noise." It may be that the intellectual tension created by his extreme intellect triggered so much creativity that his life was prematurely shortened by pure emotional exhaustion. Mozart was astonishingly immature, especially when judged to where he was in degree of sophistication with his musical compositions. It is as if his creative "genius" was so inclusive that his peer relations and behavior never were allowed to mature. It may have had a lot to do with his very over productive childhood.

Mozart was a genius with very high energy and a primary focus on composition and to a lesser extent on performance. There have been many neurological disorders attributed as origins for his behavior, including Tourette's syndrome, autistic disorder, Asperger's syndrome, attention deficit hyperactivity disorder, obsessive-compulsive disorder, and various psychiatric disorders, but again, the answer is most likely his way of dealing with the intense internal tension created by his accelerated intellectual abilities, which required a dramatic release to maintain some semblance of internal calmness (Ashoori and Jankovic 2007).

MOZART: FEATURES OF COMMONALITY AND DISSIMILARITY WITH OTHERS WITH ACCELERATED COGNITIVE ABILITY

Common features include the lack of genealogical origins for ECA but a series of early environmental modifiers or exposures that contributed to his intellectual growth. He had a very early age of awareness, age three, with an extreme musical focus at the expense of everything else in his life. Psychologically, he remained critical, intimidating, and inept socially, with all perhaps a byproduct of extreme internally directed intellectual processing. He was uncomfortable attempting to interact in society given his "genius," which is a common feature for persons with ECA.

His health remained poor, and his life was short, most likely consistent with the severe internal pressures he lived with on a daily basis. There certainly was a striking mismatch between the depth of his internal creativity and the character of his social interactions. This is not uncommon for a genius, but seems especially extreme as it applies to Mozart. It is difficult to point out dissimilarities from other persons of ECA, as he may be the archetypical "genius" from a biographic standpoint.

MOZART AND HIS LEGACY

His musical legacy speaks for itself in terms of its survivability, the absolute numbers of his total compositions, and his bedrock position in classical musical culture to this day. This is more impressive when the magnitude of his creativity is laid alongside the intensity of his many hardships, to say nothing of his very short life. One may ask why he did all that he did so intensely with a very short life, as one can with any person of genius. In Mozart's case, the degree of internal tension created by his inherent cerebral energy was so intense that he had no choice but to release in a creativity methodology to reduce the internal tension that was his constant companion. It may be that the unique DNA alignment, emergenesis, responsible for the ECA also designs an inherent need to be personally highly productive in intellectual pursuits despite the lack of any material gain or recognition.

"No one feels another's grief; no one understands another's joy. People imagine they can reach one another. In reality they only pass each other by."
Franz Schubert

"When I wished to sing of love, it turned to sorrow and when I wished to sing of sorrow, it was transformed for me into love."
Franz Schubert

"The moment is supreme."
Franz Schubert

Franz Peter Schubert

Introduction

The life of Franz Peter Schubert was as remarkable as his legacy of music has been to all subsequent generations. He was a true genius, who very early in life recognized his gifts and thankfully was allowed his creative urges to come to life despite all the pressures designed to lead him away from composing music. He rejected the normal, common, safe, and secure approach to life and chose at tremendous personal costs to himself to listen to his inner voice and create music.

The origin of this astonishing gift remains a mystery but resulted in more than a thousand compositions of beauty, breadth, tenderness, and love that continue to be a source of inspiration for many today. These compositions included songs, symphonies, sonatas, quartets, and quintets that were created in a life of only thirty-one years. He may have had no choice but to create, given the intensity of his inner tension. Why his direction or focus was music composition and not other artistic pursuits, such as writing, painting, or science, remains unknown but may be related to his early exposure to music at a very impressionable age in his home from his father. There are no apparent genetic links to family members with compositional skills or high intellectual acumen.

He sacrificed all personal comforts and worldly gains to respond to his drive to compose music. He had a circle of friends who supported him and provided some worldly solace as he persistently responded to this drive. As is true for many persons with extraordinary cognitive ability, there seemed to be a vast disconnect between the depths, mood, and power of his creations when compared to his daily social interactions. As the greatest songwriter that the world has ever known, Schubert was very shy and not verbally communicative. His life was short and difficult, but his creations live on as powerful legacies of his genius.

Schubert was committed to what he must do for personal peace and turned his back on any personal return other than the rewards of being creative. In his book, *The Unfinished Symphony, A Story Life of Franz Schubert* David Ewen addressed the issue of Schubert's genius in the following manner. "Genius—that is the only explanation that can be offered for Franz's prodigious musical knowledge at the age of ten. Genius, the fingerprint of God on man, opened the boy's eyes and mind, and showed him the naked soul of beauty; under its magic spell, the child caught a glimpse of the eternal and the infinite. Genius, God's baptismal gift to Franz, was his greatest teacher. Genius carried him to the creation of his imperishable art" (Ewen, 1931).

BIOGRAPHY

Franz Peter Schubert, nicknamed "Schwammerl" or little mushroom, was born in the Lichtensthal district, a suburb of Vienna, Austria, on January 31, 1797, at 1:30 p.m. He was the twelfth of fourteen children of Franz Theodur Schubert and his wife, Elizabeth. His father was born in 1763 in Neurdorf, near Alstadt in Moravia, to a farming family, but early on joined his older brother, Carl Schubert, in Vienna to become a school-master. His mother and father did not have any obvious creative gifts; Elizabeth was a cook by trade, and the senior Franz was a very devoted and hardworking schoolmaster. Importantly, however, Franz Theodur Schubert introduced music to the family setting very early on and played the violoncello. His brother, Ignaz Schubert, played the piano and was an early musical tutor to Franz, as was his father. His early primary tutor was Michael Holz, organist and choirmaster at the Lichtenthal parish church. The family, understandably, was only motivated to make a living and turned a deaf ear to the concept of individual creativity. He was told by his father that "teaching yielded a living and music composition a living death." Prodigious musical ability was already apparent in young Franz by the age of ten, even with these contrary pressures. He was aware of his special gift before this age and soon wanted to apply all of his energies to music composition.

At the age of eleven, Schubert attended the Imperial Convent at Universitatsplatz in Vienna and remained there for four years, from 1808 to 1812. This was a prep school for the University of Vienna. He sang in the choir at the school, which was directed by the well-known Austrian composer, Salieri. Schubert was shy, kept to himself, and began to compose music in secret. Joseph von Spaun was an early friend at the school. He was at the school in March 1812 when French cannon shots almost destroyed it.

His first musical work was the *Death Fantasia for Violin and Choir* in May 1810. His first song was *Hagars Klage*. His mother died on May 28, 1812. He remained poor, shy, and uncommunicative and only lived for his music composition. He lived an inner meditative life with rare self-expression other than his music composition. During 1812, nine works of a brilliant nature, including an overture, two string quartets, a sonata, and a quartet overture sprung into existence.

The year following he composed twenty-one works of different natures, including a cantata for his father's birthday and his first symphony to celebrate Principal Lang's birthday on October 28 (Flower 1928). In 1813, he left the academy and attended St. Anna, avoiding military service by agreeing to teach school, which he greatly disliked. He continued to teach for three more years at Saulengasse. The Mass in F was composed in 1814 during free time from teaching pursuits, but he was quite aware of the time he knew he was wasting teaching, which interfered with his focus on musical composition.

The Schubertiadem cycle, Schubert's "inner circle" of friends, was born about this time, with intellectual and revolutionary discussions in addition to many enjoyable music evenings spent together. Early members other than Schubert and his family included Spaun, Schober, Johann Mayrhofer (poet), and Professor Walterouth.

In 1815, Schubert wrote *Gretchen Spinirade D118* and *Eclkonig D 325*. He was fast becoming the greatest song composer that the world had ever known. He loved Goethe's poems and put more than seventy-five of them to song. He composed 150 songs in 1815, even while continuing to teach, and in 1816 quit teaching and moved in with Schober at age nineteen. Both Spaun and Schober recognized his genius and were aware of the destructive nature of his time away from composing that was teaching. Schober also introduced him to the famous baritone, Vogl, who became an admirer of his works and a close friend. He had a temporary romantic interest in a young soprano, Terresa Grob, who was only sixteen at the time.

This period of his life was very difficult as he was not pursuing a teaching career as his father wanted him to do, and he was not recognized for his art that he was spending so much time at, other than by his close friends. He had periods of severe depression and moods of intense religious feelings. Vogl in his diary of Schubert's songs wrote that "These were truly divine inspirations; these expressions of musical clairvoyance" and went on to comment that many of Schubert's compositions were the result of clairvoyance. (Flower 1928)

In 1818, Schubert accepted the position of musical tutor at Telise at the Esterhazy estate. He was to tutor the daughters, including Caroline Esterhazy, whom he became romantically interested in. The attraction was not mutual. He missed Vienna while at Telise in the summer and returned later in the year, November 1818, to live with Mayrhofer. In 1819, Schubert and Vogl celebrated the spring and went together to Steyr, and then to Linz and on to Saltsburg.

During this time, he would find ways to create which often included periods of rest when he would select a wild, uncultivated spot where he could sit and feel utterly alone, a spot covered with rampart growth and rich with grass and flowers. There, hidden by overhanging trees far away from the world of nervous activity, he went to work (Ewen 1931).

Later, members of the Schubertiadem included Sonnleithner, who financed initial composition publications of Schubert's; Anselon and Joseph Huttenbrenner; Umlauff; and several members of the Schubert family. Schubert greatly admired Bach, Handel, Haydn, Mozart, and Beethoven. "In Beethoven's music, Franz found a duplication of his own outlook in life. The dissonance of the *Eroica*, the unhappy voices of fate in the Fifth Symphony and in the *Appassionata*; the exuberant joy in nature of the

Pastoral—all these things Schubert understood instinctively and jarred his soul with misfortune until he, too, felt like crying out in dominant second discords like Beethoven's. And in nature, Franz, like Beethoven, found his escape from life and fate" (Ewen 1931).

In 1822, he developed early manifestations of venereal disease and became quite ill in early 1823. He was treated with mercury, lost his hair, acquired a wig, and suffered from extreme headaches and depression. He remained sick for the last six years of his life (1822–1828). "Picture to yourself a man whose health can never be reestablished, who from sheer despair makes matters worse instead of better; picture to yourself, I say, a man whose brilliant hopes have come to nothing" (Flower 1928).

Despite his unhappy state, he continued to compose brilliant masterpieces, such as his piano sonata, *Deutch 784,* in February 1823. He also composed many other pieces that year and did finally meet both Beethoven and Weber. He never was a success at opera composition, primarily because of unfortunate librettos. It was a great frustration for him because opera composition was looked on most favorably by his contemporaries. The depth of his periodic depression can be appreciated in his letter to Leopold Kupelwieser during this period. "Every night when I go to bed, I hope that I may never wake again, and every morning renews my grief. I live without pleasure or friends. Schwind pays me an occasional visit" (Flower 1928).

Schubert returned to Telise in May 1824 but quickly discovered that his earlier perceived romance with Caroline Esterhazy could not be rekindled. He remained there, however, but continued unhappy. He wrote, "Here I sit now, all alone in farthest Hungary, where I unfortunately allowed myself to be enticed for the second time, without having even a single person to exchange a clever word with" (Neumayr, 1994). He returned to his old haunts in Vienna in 1825 and discovered his lost gaiety, creating thirty compositions. His health had improved somewhat by July 1825. He had a passion for Goethe and set as many as seventy of his compositions to song. Schubert was rebuffed by him, only to be appreciated by him after his death.

Schubert certainly experienced profound grief throughout his life. Perhaps, as he wrote, this was important for his creativity. "Grief," he wrote bravely in his diary, "sharpens the understanding and strengthens the soul, whereas joy seldom troubles itself about the former and makes the latter either effeminate or frivolous" (Ewen 1931). His religious devotion was expressed in music during these years; the *Ave Maria* being the crowning achievement. "They were greatly surprised at my piety, which I expressed in a hymn to the Virgin, (the *Ave Maria*), which apparently moved everybody" (Flower 1928).

Diabelli, the music publisher, introduced Schubert to Beethoven, whom he idealized from a far. Beethoven's observation was that "there was something heroic about a man who could compose so vast an amount of music in the face of the world's indifference." The sheer courage of the man impressed Beethoven (Ewen 1931). Beethoven died on March 27, 1827, and Schubert wanted to write a requiem but was unable. Schubert's total life's earnings were L575 for a thousand songs, symphonies, operas, sonatas, concertos, dances, etc. He considered himself a total failure, at least as judged by the standard measurements of the day, and was obsessed with his failure up to his death.

He gave a concert on March 27, 1828, one year after Beethoven's death, which led to a flood of new creativity. He became ill again after the concert and moved to his brother's house in September 1828 in Neue Wieden (Ferdenhand), a suburb of Vienna. He died of typhus on November 19, 1828. "Isn't there some consolation in knowing that one's body may rot in the grave but the fruit of one's spirit will live forever?" (Ewen 1931)

There lay the supreme tragedy of his tragic life. He was denied not the innate talent necessary for the achievement, but the time in which to accomplish and to prove it (Goulding 1992).

SCHUBERT AND ACCELERATED COGNITIVE ABILITY

He recognized his gift early in life and immediately focused on creating music. He wanted to compose music at the cost of all else, despite its impractical nature. Why did he have this intense passion for music composition? Is the accelerated basic musical acumen directly inherited via the neurobiological template along with the ECA, or does the musical focus only occur because of the nature of his early exposures as a significant modifier? This remains a mystery. Does the gift of genius only apply to musical composition for Schubert, or could he have applied his genius in other realms? He was not a conductor or instrumentalist. He did not he want to be a teacher. There was some musical stimulation with his father and older brothers, who were both musical, but there are no dramatic examples of compositional or musical skills in the family lineage. The sheer power of genius has not been identified elsewhere in his genealogy.

Schubert is so loved for his gift of music. Everyone appreciates that he focused his ability and felt he left his legacy for all us to enjoy despite the hardships it created for him personally. The sheer force or power of will to continue in such a prodigious fashion is astonishing. Is this energy a common feature of genius, demanding an outlet, and in this situation, a positive focus? This force is consuming and demands intense focus. It overrides all other lesser concerns and perhaps accounts for the common observation that most geniuses are eccentric with borderline psychopathology. He created because he had to modify or partially relieve the intense inner tensions that he felt.

No genetic, environmental, educational, or experiential clues exist for Schubert's genius, but perhaps the core element is his early recognition of his gift and his demands to utilize it in a positive fashion. There may be many persons of genius, but the difference may be the early recognition, the ability to focus this extreme energy and identifying the avenue for expression and ignoring all else that tends to dilute the intensity of the focus. This may be a key insight to the genius process, at least as it applies to Schubert in the early recognition of the energy and specific focus along with intense devotion to it despite competing pressures. He certainly found a very positive focus that resulted in his wonderful legacy of music for subsequent generations.

Many geniuses may not recognize their gifts and become emotionally ill because of the tension created by the heightened energy. Thus, they may not utilize their gifts in a positive fashion and even become destructive with their minds and lives. How crucial was his father's and brothers' musical ability, although modest, in directing Schubert to compose music and utilize his genius in a very positive and focused way? What role does the lack of comfort and general misery of his daily life play in his creativity? Are emotional stress and strain and the lack of comfort essential and acting as drivers of creativity? He certainly was uncomfortable most of his life and had essentially no emotional or compensational rewards for his work.

We do not know what the core anatomical and physiological framework necessary for his genius was, other than the sheer power of his intellect and his demand to focus at a very early age and throughout his life. He did lose a parent early in life, his mother, when he was fifteen. His father was somewhat difficult because of his objection to his son's focus. Schubert rejected the material returns that were available with teaching, which was the family profession.

The role of intellectual stimulation, the Schubertiadem in his creativity, is interesting and may have provided the staying power crucial for its success. The role of illness, that is to say infection, depression, the lack of any successful romantic relationships, and the sensitivities that this creates may have contributed as well.

SCHUBERT: FEATURES OF COMMONALITY AND DISSIMILARITY WITH OTHERS WITH ACCELERATED COGNITIVE ABILITY

Schubert's life was very typical, if not the mold, for individuals with extraordinary cognitive ability. His life had many of the same attributes of other individuals with this cognitive energy, including very early exposure (music) to his area of intense focus and early awareness of his gift as a passion. He was generally shy and not outgoing and experienced the loss of a parent early in his life. He had precocious involvement in his focus and applied himself with tremendous energy, teaching most of the day and composing music late into the night. He had awareness and admiration for his predecessors, a love for nature, and a need to spend much time alone to feed his creative needs. In addition, there were chronic medical problems, a clear mood disorder, a very early death, a marked disconnect between the depth and beauty of his creations compared with his societal interaction, and a complete lack of material needs or positional gains in the society in which he found himself.

Areas of dissimilarities are few but might include a severe lack of self-confidence; needing a strong, supportive group of peers, which was very important to him; perhaps an earlier than usual recognition of his gift, and a focus with an intense need to compose at the expense of all else.

SCHUBERT AND HIS LEGACY

Schubert in his own words, best describes for him the common association of genius and emotional pain. "My music is the production of my genius and my misery." Schubert composed more than a thousand compositions in his short life of thirty-one years, including songs, symphonies, sonatas, quartets, and quintets, as mentioned above. He received little compensation and essentially no recognition during his life for all these marvelous works. He loved poetry, and many of his lieder were stimulated by poetry, especially those of Goethe. He was master of the sonata and symphony and was the world's greatest creator of lieder, composing over five hundred pieces including the beautiful *Ave Maria, Gretchen at the Spinning Wheel, the Erlkonig,* and the *Die Winteresise* cycle." He wrote nine symphonies, but there are only sketches of number seven in existence. His most admired symphonies were the great C Major Symphony No. Nine, No. Five in B Flat, and the unfinished Symphony No. Eight in B Minor. The ninth symphony was composed in his last year and was on the plane of Beethoven's choral Symphony No. Nine, but most think that he never heard the Beethoven's ninth played but may have seen the score. Schubert wrote more than twenty quartets, several trios, and several piano trios, the best known of which was his Piano Trio in B Flat, Op. 99. His most famous quintet is the "Trout" Quintet in A for piano and strings. His piano music is astonishingly beautiful and includes the inspirational Fantasy in C, Opus 99 D760, the Piano Sonata in B Flat, D960, the Piano Sonata in A Minor, D845, and multiple impromptus, including the three impromptus, D946, Mass in E Flat, and his Requiem.

These compositions remain permanent treasures for all future generations as his gifts to the world, created by his genius and secondary only to his remarkable focus. He accomplished all this despite the lack of any kind of recognition by his contemporaries or monetary reward during his short life. He knew he had this gift of creation and was relentless in his pursuit of musical composition, despite the contemporary focus attempting to lead him away.

The music is astonishing in its beauty, breadth, power, and romantic sensitivity. He was providing the true bridge between the classics and the romantic movements. His legacy has and will continue to provide wonderful moments of insight and joy for all who listen and understand it.

The importance of music as a way to better understand self and life is apparent in exquisite detail in Schubert's multiple musical compositions. It is as if the divine provides another way for us to better understand the beauty of life and the great riches of inner peace and understanding that come from appreciating and understanding great music. Music provides moments of great and beautiful insights that we all can receive, as is apparent from these gifts of Schubert.

The beauty and power of his music is astonishing. One can receive the great emotions of pain, love, strife, depression, exhilaration, tragedy, etc. conveyed as a musical story or painting that becomes more electric the more familiar it becomes. He had a great love for poetry and was the master of the Lieder (or song), transferring the literature into music. Interesting, however, was his failure with opera, which may have been a problem with the libretto he attempted to deal with. His concertos and sonatas are without equal, as is his chamber music. His last two symphonies remain exceedingly popular, and there is palpable pathos in his requiem. For one man to create this much emotion in art form as music is hard to believe, but it may have to do with his intellectual capacity coupled with severe personal emotional strife and tragedy.

Do creative people who are geniuses by definition live a life of personal pain that is central to their creative outputs? Schubert sacrificed all to create his music, and by most standards, lived a very painful and far too short of personal life. It may be that too much comfort and satisfaction personally is incompatible with significant artistic achievement. His music is highly emotional and paints a picture that becomes more understandable and enjoyable the more it is heard and studied.

"What lies behind you and what lies in front of you, pales in comparison to what lies inside of you"
Ralph Waldo Emerson

"To believe your own thought; that is genius."
Ralph Waldo Emerson

"Do not go where the path may lead, go instead where there is no path and leave a trail."
Ralph Waldo Emerson

Ralph Waldo Emerson

Introduction

The origins of Ralph Waldo Emerson's extraordinary intellectual prowess and creativity may be clearer than most other intellectual gifted individuals, given the extent of the lineage information now available. It clearly displays the remarkably heightened intellectual curiosity that was so prevalent in multiple Emerson family members in the generations preceding him. He may well serve as an important exception of the argued central importance of emergenic genetic influences in the development of extraordinary cognitive ability as discussed in the section on basic neuroscience.

Multiple family members from several generations were very intellectually curious and most were devoted to promoting knowledge and scholarship. Standing behind Ralph Waldo Emerson's cradle might have been described the shades of eight generations of ministers or other bookish men, introspective and dreaming of another and better world, and married to devout, indoor women (Russel 1929). Mary Moody Emerson, Ralph Waldo Emerson's aunt, was the intellectual leader of the family and quickly became Emerson's model to follow at a very early age. She remained the key intellectual influence on him the remainder of her life.

He struggled early in terms of defining his intellectual focus with initial exploration and subsequent disillusionment as a member of the ministry, coupled with the untimely death of his first wife, resulting in a prolonged period of reflection and ultimate major redirection of his intellectual energy. During this reflection, he was exposed to new ideas and friendships that, combined with his voracious reading habits, set him off on a new intellectual path that would be his life's work. His immediate environmental exposures or key modifiers, always an area of great interest in intellectually advantaged individuals, along with multiple, important contacts made on his initial trip to Europe, were obviously important to the initial nurturing and direction of his intellectual energies.

The Environs is always a key issue of discovery with extraordinary cognitive ability, but Emerson's story may serve as a prime example for how crucial this factor may be in the overall direction that accelerated intellectual energy takes. His writings and lectures soon became the spark for a revolution in thought and literature in America, an intellectual independence from Europe that was central to the growth of American thought. He was the key figure and leader of the transcendental movement in America, which continues to have a great impact on intellectual thought in America and throughout the world to this day. Uncharacteristically for persons with ECA, he lived a long life that was compromised late by advancing dementia with memory loss. His intellectual legacy was remarkable and remains a wonderful gift for the ages for anyone with questions about man's existence and his place in the world.

BIOGRAPHY

Ralph Waldo Emerson, consistent with his underlying philosophy of the potential primacy of one's own home environs, did not stray far from his place of birth his entire life, except for several important travels to Europe and the Western US. He lived in the Boston and Concord areas of New England. He was born in Boston on May 25, 1803, on Summer Street near

Boston Commons, the third of six children to Reverend William Emerson and Ruth Haskins. His father was in the twelfth year of his pastorate at the First Church of Boston and died when Ralph was only eight in 1811. He was raised and educated by his mother and aunt, Mary Moody Emerson, with very limited family monetary resources.

Mary Moody Emerson was the single most important mentor for his early, unique intellectual curiosity and thus was a key element in his early education. She also provided crucial early intellectual stimulation to all his brothers—William, Edward, Robert, and Charles. A sister lived only until 1814. Ralph Waldo Emerson was very intellectually curious from early in his life and read widely and voraciously. "His voice was slow and musical, and occasionally when he spoke, there was a flash in his expression that vaguely suggested some strange inner power" (Brooks 1932).

There was an unfortunate, strong presence of tuberculosis in the family, and his brother Charles died from the disease in 1836. Edward was very bright but had a "mental collapse" at age twenty-nine and died in 1834. Robert was born in 1807 and was retarded.

Early on, Emerson wanted to be a poet, an orator, a minister, or a writer. He started his "commonplace book" at a very early age and thus began accumulating quotations from his readings plus writing his own personal observations on a very frequent basis. He had an enormous appetite for knowledge and read everything that he could get his hands on. He had an insatiable curiosity, was keenly aware of his surroundings, and took in everything. He attended Harvard at age fourteen and obtained his degree at age eighteen and was considered an average student. He briefly taught school at William's

School for Women after leaving Harvard from 1821 to 1823 but soon found that this was not of great interest to him. In December 1823, at age twenty, he had a sudden awareness of self-validation and wrote, "Who is he that shall control me? Why may not I act and speak and write and think with entire freedom? What am I to the universe or, the universe, what is it to me? Who hath forged the chains of wrong and right, of opinion and custom? And must I wear them" (Richardson 1995)?

He was a voracious reader all of his life and during these early years, studied Plato, multiple German philosophers, Montaigne, Rousseau, Emile, and Samuel Reed. His idol at Harvard was Edward Everett, who taught Greek literature.

He became engaged to Ellen Tucker on December 17, 1828, in Concord, Massachusetts, and after graduating from seminary school, started his career as a Unitarian minister, replacing Henry Ware at the Second Church in Boston. He was ordained in March 1829. He married Ellen Tucker September 30, 1829, in Concord, New Hampshire. However, Ellen soon became very ill with a progressive illness in 1830 and died on February 8, 1831. About this time, he began to move away from an early focus on religion toward science and thus a fundamental philosophical realignment was greatly enhanced with the emotional turmoil given the untimely death of his wife. After a prolonged period of reflection following her death and his separation from the church, he traveled to Europe for twelve months, leaving December 10, 1832. After spending five months in Italy, he went to Paris where he was exposed to Jarden des Plantes. He later went to England, where he had a very fortunate opportunity to meet and converse with Wordsworth, Coleridge, and Thomas Carlyle. All these contacts came at a very fortunate time of great receptivity for Emerson and proved to be a monumental influence on the direction and energy of his subsequent intellectual pursuits. He returned to Boston on October 9, 1833, and began his first great essay, "Nature." Emerson expanded his thoughts on self-reliance and discovered the importance of his journals as source material for formalizing his new ideas in essays and lectures. American transcendentalism began with Emerson with his lecturing and essays in 1834.

At this time, he began to correspond with Thomas Carlyle in England and was greatly influenced by Goethe's works. He always loved poetry and began to write his own poems. "Rhordora" was written in May 1834 and "The Snow Storm" at about the same time. He lived at the Old Manse with his step grandfather, Ezra Ripley, in Concord Massachusetts, and with his brother Charles and Mother Ruth. He developed a great interest in biography, and in 1835, at age thirty-two, he developed a series of biography lectures that seemed to encourage him about his own potential abilities as an individual.

He married Lidian Jackson in September 1835 and moved to his new home in Concord, Massachusetts, where he lived the rest of his life. In 1836, he began his book on nature. Emerson, while on a walk in March of that year, developed the idea for his book with the following quote. "As I walked in the woods, I felt what I often feel, that nothing can befall me in life, no calamity, no disgrace (leaving me my eyes) to which Nature will not offer a sweet consolation. Standing on the bare ground, with my head bathed by the blithe air, and uplifted into infinite space, I become happy in my universal revelations. The name of the nearest friend sounds then foreign and accidental. I am the heir of unaccustomed beauty and power" (Richardson 1995). He continued to read voraciously, focusing on the works of Goethe, Carlyle, Coleridge, and Boehme.

The Transcendental Club was formed on September 8, 1836, as a protest against the "arid intellectual climate" of Harvard and as a forum for new ideas. The original members included Henry Hedge, George Putman, George Ripley, Margaret Fuller, and Emerson. The members of the club were responsible for the first publication of the *Dial* in 1840, a collection of new ideas by the members of the club. At this time of writing "Nature" and its subsequent publication, he was working very hard with multiple lecture preparations, prolific expansion of his "common place journal," and composition of his own ideas in essay form.

He began to show features of a manic/depressive mood disorder, with periods of hypomania resulting in massive outputs of both ideas and written material. Henry David Thoreau became a disciple of his after the publication of "Nature" in 1837 but soon became quite independent with his own concept of self-reliance. Both Emerson and Thoreau kept elaborate notebooks to keep track of material and to treasure their moments of illumination and insight (Richardson 1995).

"The Divinity School Address" on July 15, 1837, attacked the then-present day formal historical Christian tradition by declaring the presence of a divine nature in all individuals. He was then cut off from Harvard after his "Divinity School Address," but he continued to study in wide areas, reading multiple areas of interest. On a day in August of 1837, in the old meetinghouse of the First Parish of Cambridge, he delivered the annual oration of the Phi Beta Kappa Society, an essay called "The American Scholar." This was later described as the Declaration of America's intellectual independence (Brooks 1932). By 1839, he had developed a book of essays written from his journal entries. His journal was all indexed to a four hundred-page master index book, plus he had a large biographical index with 839 men and woman with specific quotes.

This two and a half years was an obvious time of high energy for him, likely reflecting a period of hypomania as part of his manic-depressive mood disorder. His fundamental conviction was of the essential oneness of all things. He found his power through absolute self-reliance.

His essays were published in January 1841, and a period of depression followed for a time. His essay themes included character, manners, art, politics, and friendship. He became the editor of the *Dial* in 1842, replacing Margaret Fuller.

He continued in this capacity for the next two years. Perhaps more important , he spent many hours walking with Thoreau, Ellery, Channing, Caroline Sturgis, and Margaret Fuller, expanding and solidifying his transcendental ideas with his colleagues. Richardson comments that "At almost any period of his adult life, Emerson seemed to have periods of half epiphany and half cordwood. He needed both ecstatic experience and pie for breakfast" (Richardson 1995). He gave up editing the *Dial* in 1844 as he was upset with the general lack of support for the publication. In 1844, he gave his Concord speech concerning antislavery issues and began to work for the abolition of slavery, moving from an observer to an activist. Emerson hated to waste time away from reading and creating and found correcting proofs of his new volume of essays a very laborious task. *Representative Men* was written in 1844, a modern nineteenth century rendition of the ancient Plutarch's lives. He wrote about Plato, Swedenborg, Montaigne, Shakespeare, Napoleon, and Goethe, which represented for Emerson the full flowering of some single aspect of our common nature.

He traveled from one small town to another by train, lecturing about great men of the past. He lectured at the Lyceum, not the university, as he was not invited back to Harvard following his divinity school address until after the Civil War. In 1846, he continued to compose poetry and greatly enjoyed studying Persian lyric poets. He also had periods of restlessness and dissatisfaction after the publication of several of his books, coupled with Lidian's illness, difficult finances, and clinical depression. "No news or visitor from abroad, no lion roars, no mouse chirps, we have discovered no new book, but the old atrophy inanition and drying up proceeds at an accelerated rate" Reflecting the clinical depression he had at the time. (Richardson 1995).

He had an endless effort to self-reform as Samuel Johnson and was restless, depressed, unfulfilled, and self critical—all typical of persons of great cognitive ability. He decided to go back to Europe and boarded the packet "Washington Irving" for Liverpool. He gave numerous lectures while in England—sixty-seven in total—and he also met individually with Thackeray, Thomas Carlyle, Tennyson, Macaulay, Disarili, Palmerston, Wordsworth, Dickens, George Eliot, Mathew Arnold, and Chopin.

Despite these travels, Emerson remained depressed and moody. His relationship with Thoreau deteriorated. Brooks spoke to the depth of Emerson's depressions with the following quote. "He was like the maple trees in the spring when the sugar flows so fast that one cannot get enough tubs to contain it. And then came hours of pain, sterility, and ennui, and he sat out the day and returned to the necessities of the household, doubting if all this waste could ever be justified." (Brooks 1932).

The depression lifted in 1850 with a new burst of energy and a reexamination of all his philosophical concepts. He visited many cities in the eastern United States and took a trip West, especially admiring the independent mind-set of the individuals he met there. He wrote *The Conduct of Life*" proceeding with new themes of thought. Emerson became more active in the antislavery movement after the passage of the Fugitive Slave Act, becoming very angry with Webster for supporting it in Congress. He thought that no law had any validity that was contrary to the law of Nature (Brooks 1932).

In 1853, at the age of fifty, he became aware of diminishing energy levels. He continued to lecture frequently but wrote less. He gave seventy-three lectures from October 1854 to September1855, as an example. In 1867, he gave the annual address at the Harvard Chapter of Phi Beta Kappa but became frustrated when he could not see his papers well. He then became embarrassed by his failing memory and continued to lecture but less frequently. One can only imagine the striking intellectual changes his peers must have became aware of during this period of his life and energy decline. He died of pneumonia in 1882 at age seventy-nine,

EMERSON AND ACCELERATED COGNITIVE ABILITY

Ralph Waldo Emerson clearly functioned all his life at a very high cognitive level, which was evident from very early in life with his extraordinary intellectual curiosity. His memory was extraordinary, with almost complete recall from the books that he read throughout his life. His aunt, May Moody Emerson, who was intellectually gifted herself, recognized very early on the intellectual prowess that Emerson displayed and help to mold it. He came from a family of persons with extraordinary cognitive ability with direct inheritance of this trait rather than the need to necessarily postulate the genetic emergenic theory of trait inheritance. His may be a good example of the power of the environmental markers in molding not only the direction but also the intensity of the chosen focus of intellectual activity in combination with a clear genetic predisposition for ECA.

With the strong supportive environs, he began to energetically pursue intellectual areas of concern, perhaps further stimulated by the combination of his disenchantment with dogmatic religious thinking and the personnel emotional crises after the early death of his first wife. He began to create poetry and utilize his "common place book" to write his own thoughts in the form of essays about various topics along with giving multiple lectures. He remained keenly interested in all that he could learn and read aggressively all his life in many different venues and from many different cultures. He continued to fuel his intellectual energy with the close relationships of peers, and in concert with them, he formed a forum discussion group and helped publish the *Dial*, a publication of their numerous essays.

He had a manic/depressive mood disorder, common in persons with extraordinary cognitive ability. He avoided chronic illness and lived a very long life for seventy-nine years, but was compromised in the later years by an unfortunate and slowly progressive dementia.

EMERSON: FEATURES OF COMMONALITY AND DISSIMILARITY WITH OTHERS WITH ACCELERATED COGNITIVE ABILITY

Common features include the very early appearance of a unique curiosity for all that he met. It was encouraged by his peers and family members, who also were intellectually gifted. He also had the clear manic/depressive disorder with secondary behavioral correlates so common in persons of accelerated cognitive ability. Inheritable traits also played a very important part, as his family for generations preceding him had many individuals with similar accelerated intellectual capacity. He was fortunate to focus relatively early in his life on his intellectual pursuits and was blessed with a supportive and stimulating environment composed of peers and other family members who continued to encourage him. Unlike others with extraordinary intellectual achievements, he lived many years, to age seventy-nine, uncompromised by recurrent illness save for his very unfortunate evolving dementia later in life, which greatly compromised his ability to continue to create or even lecture. He was also very comfortable emotionally almost his entire life except for bouts of depression and except for the dementia that appeared late. This dementia, by its degree and late onset, is relatively unusual for highly intellectual persons. One also might argue that a clear difference was a more common Mendelian genetic explanation for his intellectual energies, given what we know of his family tree, as opposed to the more usual need to implicate an emergenic genetic phenomenon to explain the unique occurrence of a neurobiological framework critical for this level of accelerated cognitive ability.

EMERSON AND HIS LEGACY

Emerson was the dean of American writers, philosophers, poetry, and academic lectures. He was responsible for the intellectual independence of

America in the 1830s, with his numerous essays and lectures based on his new concept of man and his relationship to Nature. His transcendental ideas were radical for the day and were highly criticized by many but remain today as a very important concept that has stimulated thought and action for many in later generations. He promoted individual intellectual independence with a keen sense of self–reliance, which is the basic core of his transcendental ideas. He promoted the transcendental ideology that continues to flourish today, and his ideas have impacted many generations of thinkers. His books and lectures continue to influence modern thinking and have had an impact on religious thinking, especially Unitarian theology.

The concepts of self-reliance, our oneness with nature, and intellectual independence that make up some of the core elements of transcendental ideology burst on the stage of American literature with Emerson's intellectual energy and are reflected in his writings and many lectures. Today, we are very fortunate to be able to study and further explore his ideas, which are just as germane now, perhaps even more germane, to our lives and culture than they were in his day.

"Simplicity is the final achievement. After one has played a vast quantity of notes and more notes, it is simplicity that emerges as the crowning reward of art."
Frederic Chopin

"Sometimes I can only groan, and suffer, and pour out my despair at the piano!"
Frederic Chopin

Frederic Chopin FRÉDÉRIC FRANÇOIS CHOPIN

INTRODUCTION

Chopin, the apotheosis of the piano, was born in humble surroundings near Warsaw, Poland to a family of musicians. He was clearly precocious, and it was recognizable at a very early age. He was fortunate to have a very supportive family emotionally and intellectually. Exposed to music from his earliest childhood years, he responded enthusiastically with obvious and remarkable abilities to learn rapidly and create his own musical compositions. It is not entirely clear why Chopin's early and only focus in life was music, but it may well have been in part the importance of his very early musical exposure in the home, with his parents interested in singing and playing various musical instruments.

One of the major tragedies of his life was his severely compromised health, which included pulmonary tuberculosis, "rheumatism," gastric disorders, headache, and poor dentition. Manic-depressive disease was also a major problem for him his entire life, which some would suggest is a common important ingredient for most intellectually advantaged persons. He was very emotional and insecure, even through he was raised in a most secure background without any obvious mental disorders in other family members.

He was aware of his special cognitive gifts early and focused all his energy to studying both musical technique and composition. The immediate central question that surfaces from these observations, which is true for all persons of extraordinary cognitive ability; is what was the prime responsible factor for their specific creative focus—music, for Chopin?

He was exposed to music from his earliest childhood days, responding to it naturally, enthusiastically, and indeed astonishingly (Szulc 1998). As with other persons of extraordinary cognitive ability, the central question is whether there was a genetic, predetermined aptitude for music ability, Or did Chopin, with all his naturally occurring intellectual gifts, focus on music solely based on his exposure to music in the home of his parents? Perhaps this was a key environmental modifier, enhanced by very capable musical tutors early in his life?

There was no clear extraordinary music ability apparent in a detailed study of his family lineage. However, this may not eliminate an inherited aptitude. This is always a very curious question for these gifted individuals. Why did Chopin choose music and not literature, art, or science? The answer to this is not at all well understood at present but may as well been the importance of music in the home environment. This is the most obvious possible explanation for this direction, given the known data. But it may not be the correct one, as there might have been extraordinary musical aptitude in past family members, unknown at present. However, as discussed in the basic science section, the specific focus may result from a unique alignment of the DNA in the genetic code, the theory of emergenesis, resulting in a specific extraordinary aptitude for music along with unique extraordinary intellectual ability. In Chopin's case, it was not dependent on previous family history of musical prowess. Many wonderful biographies are devoted to Chopin, many of which focus on his music, but there are more general biographies, including those by Murdoch (1935), Leonard (1946), Szulc (1998), and Zamoyski (1980), to name only a few. This biography will be focused on his ECA and its central role into who he was and the creation of his musical gifts.

BIOGRAPHY

Frederyk Franciszek Chopin's birth date remains controversial, but it was either February 22 or March 10, 1810. His family lived in a humble setting twenty miles west of Warsaw, Poland, in the small town of Zelazowa-Wola on the Ultrata River. He was the second of four children. His father, Nicholas, was a transplanted French tutor who played the violin and flute. His mother, Justyna, was of humble Polish nobility and had singing abilities and played the piano. They were married in 1806. Chopin lived in Zelazowa-Wola only six months when the family moved to Warsaw, where his father became a tutor at the Warsaw Lyceum.

There were four children. None were gifted, save for Chopin, but his youngest sister, Emilia, wrote poetry at the early age of eight. Chopin was obviously very precocious, with a very early age of awareness of his intellectual gift fortunately coupled with a musical focus provided for him by his family and early tutors. At age four, Frederyk's mother introduced him to the piano, and by age six, he was creating considerable interest in Warsaw by his obvious precocious abilities in piano compositions of his own creation. Adalbert Zywny, a violinist who focused on eighteenth-century music of Bach, Haydn, and Mozart, was Chopin's first tutor at age six. Chopin remained committed to the spirit of eighteenth-century music all his life and took his taste and musical inspiration from it; although he was very much a part of the new romantic age of music as reflected in his own compositions. He wrote the Polonaise in G Minor at age eight and soon became known as "Mozart's successor."

On February 24, 1818, he played a concert in Warsaw, and after it, his fame spread as he became known as the new Polish genius." Joseph Elsner, head of the Warsaw Conservatory, became his next tutor, worked with his playing technique, and helped inspire early compositions. He remained humble, likely secondary to the influence of his father, who was determined not to let his musical attainments go to his head. His father insisted on treating his son's gift as a pleasant amenity rather than the central feature of his son's life (Zamoyski 1980).

He attended school full-time at the Warsaw Lyceum. He focused primarily on music and had many friends, including his lifelong closest friend, Tytus Woyciechowski, whom he spent his summers with out in the Polish countryside. He had exposure to other great musicians in Berlin and Vienna, including List in Vienna when he was thirteen. Chopin was not timid and had developed a great sense of humor, had irreverent wit, and had a keen ability to imitate others—all common characteristics of individuals with ECA.

He never had robust health and was generally weak and frail throughout his youth. He clearly had pulmonary tuberculosis for more than thirty years, a disease known then as pulmonary consumption or phthisis. He was exposed to country folk Mazovian dances during summer holidays from school and later wrote poems in musical language in response to his emotions. He was also exposed to Polonaises, and the very special Jewish music of the small villages.

He played in the grandest drawing rooms in Europe but remained humble thanks to his father, who was a very sobering influence on him. He began to write waltzes and mazurkas at age fourteen, gave a public concert in Warsaw for the Russian Tsar, Alexander, and completed his first commercial publication, the Rondo in C Minor, Opus 1 in 1826 at age sixteen.

Following school at the Warsaw Lyceum, he decided to enter the conservatory in Warsaw in 1826, rather than go on to the university. He was at the conservatory for three years with his tutor, Joseph Elsner, and focused on musical theory and counterpoint. Elsner wrote in his own diary that Chopin, a third-year student at the conservatory, "opened a new era in piano music through his astonishing playing ability as well as through his compositions" (Zamoyski 1980).

Chopin, however, was always convinced that for him, the piano was the perfect instrument and that he could and would achieve more with the keyboard than one could with a hundred voices on an opera stage (Szulc 1998). His younger sister, Emilia, died from tuberculosis, and his health remained poor with multiple pulmonary complications from his chronic infection with tuberculosis. He saw and admired Paganini and decided to write for the piano as a solo instrument, much as Paganini had achieved for the violin. Etudes 8, 9, 10, and 11 of Opus 10 were composed following the visit of Paganini. He wanted to become the Paganini of the piano.

He went to Vienna in 1829 and thought that the musical elite of Europe were there, as the musical performances exceeded his expectations. He played in concert as well and received praise from Count Lichnovsky, who had been a close friend of Beethoven. He visited Prague, Tirpitz, and Dresden before retuning to Warsaw in September 1829. "The creative spirit of the young composer has taken the path of genius. The critics could not make up their minds as to whether it was his playing or his compositions which were the more remarkable, and comparisons with Mozart and Hummel were bandied about liberally" (Zamoyski, 1980). "He was bashful, suspicious, very romantic, and extraordinarily sensual." (Zamoyski, 1980) He lived on Kohlmarket Street in Vienna with Tytus in November 1829, visited the opera frequently, and loved the musical stimulation of the city. He composed in the morning, had tea, and took walks in the afternoon. Then he attended numerous soirees until midnight.

His friend Tytus left for Warsaw at the start of the revolution in Poland, but Chopin chose to stay on in Vienna, eventually becoming depressed, lonely, and quite introspective with much self-pity. Despite his deep depression, he composed the Scherzo in B Minor Opus 20 and the Ballade in G Minor Opus 23, which achieved depth of sound and feeling with a single and self-sufficient instrument.

He left Vienna July 20, 1831, for Linz, later going to Salzburg, Munich, and Stuttgart and eventually arriving in Paris in September 1831. This was a remarkable time for young musical "geniuses." Chopin was twenty-one, Mendelssohn was twenty-two, Schumann was twenty-one, Liszt was twenty, and both Verdi and Wagner were seventeen.

His move to Paris in September 1831, rather than returning to his homeland in Poland, was a major milestone in his life, as it became his home for the rest of his life. He never returned to Poland. When Chopin set foot in Paris in 1831, he instantly realized that indeed, he had come to the cultural and musical capital of the world. In Paris at the time, he found himself in a very comfortable environment to continue with his work. The Romantic Age for music was just beginning to peak there. He composed in and was of the Romantic

Age but continued to be closer to the spirit of the eighteenth-century classical music, from which he took his taste and his musical inspiration. It is extraordinary to realize that Berlioz, Bellini, Chopin, Liszt, and Mendelssohn, all of the same generation, born between 1803 and 1813, found themselves and each other in the same place and were devoted to musical creation. Chopin withdrew more into himself and became increasingly suspicious of any intimacy. Friends were struck by his reserve. He admired the French but was most comfortable around Polish ex-patriots. He was attracted to Franz Liszt but found him of a very different temperament. He did not fit easily into the Romantic Movement Beethoven started, but he knew all the new romantic artists in Paris. According to Zamoyski, he did not care for Schubert, Schumann, Mendelssohn, or Weber. (Zamoyski, 1980) He spent the winter 1832 and 1833 in Paris; heard Fields in concert, the great Irish piano composer whose style he was greatly attracted to; and heard Beethoven's Ninth in concert for the first time. He met, liked, and played a concert with Liszt and had a generally positive relationship with him the rest of his life, even given the great differences between their personalities.

His true destiny, Chopin was convinced, was to continue composing for the piano. It was the calling of his genius (Szulc 1998). He supported his lifestyle by giving piano lessons, as many as five per day, six days per week at twenty francs per lesson. He achieved a rather special reputation without any apparent struggle, and by the beginning of 1833, he was considered one of the brightest stars in Paris.

He continued to show great ability with improvisation and was considered the greatest piano teacher in Paris. His reputation grew as a composer in 1833 to 1834, although he was busy with lessons during the day and social and musical events in the evenings. He composed four mazurkas Opus 24, two Polonaises Opus 26, and 2 Nocturnes Opus 27. Charles Halle writes that there is "nothing to remind one that it is a human being" who produces his music. "It seems to descend from heaven, so pure and clear and spiritual" (Zamoyski 1980). He loved to improvise late into the night at soirees, but the late hours further aggravated his compromised medical conditions. He clearly had recurrent influenza symptoms because of his many late nights.

He met Mendelssohn in May 1834 at Aachen. Mendelssohn was very impressed with Chopin's piano prowess and compared him to Paganini and what he had done for the violin. He met his parents in Karlsbad in August 1834 and spent a month with them. Later, he moved to a Dresden meeting with Maria Wodzinskis, whom he was quite attracted to earlier in Poland along with the rest of her family. He proposed to her at Marienbad, but her mother said that he had to prove to her family his good health first before any consideration of marriage. He developed severe pneumonia on his return to Paris, which represented a progression of his tuberculosis. That started a gradual decline in his health, eventually resulting in his early death in 1849 at the age of thirty-nine.

This progressive illness may have produced growing psychological distortions over the years, accounting for some of his personality disorders, including manic-depressive illness. He met George Sand, a famous novelist in Paris (Aurore Dupin Diudevant) on October 24, 1836, at a reception Liszt had invited him to at the Hotel deFrance. George Sand adopted Chopin, as a mother would a child, and remained a central person in his life for many years.

He was very creative during this time, with compositions including the Nocturne in B Major, Scherzo no. 2 in B-Flat Minor, Ballade no. 2 in F Major, and the Sonata no. 2 in B-Flat Minor, but his heath continued to deteriorate. He visited London in July 1837, returned to Paris after three weeks, and continued to be very creative during this time. On his return to Paris, his relationship with George Sands began in earnest in the spring of 1838 and lasted for nine years until two years before his death.

He continued to be quite creative despite his deteriorating health and his new relationship with George Sand. He had great powers of concentration, liked to be alone, and had a wealth of inner emotion that did not need external means to stimulate him and lead to musical creations. He planned a trip to Majorca in late October of 1838 with George Sand and her family, which was a very difficult experience for him given the winter conditions. He was still able to compose some of his more memorable compositions there. He had manic-depressive crises with hallucinations during this stay at Majorca but survived the trip and returned to Paris. It was nearly a miracle that Chopin, at age twenty-nine, was alive when he returned from Majorca. He was able to summon reserves of energy to go on surviving and creating his wonderful music (Szulc 1988).

He continued to sustain himself economically by giving piano lessons, which he did in the mornings and part of the afternoon which he described as a "mill" for his lucrative hours. He was most productive at Nohant in the summer, where he spent time with George Sand and her family. At Nohant he composed the Scherzo and Finale of the B Flat Minor Sonata, Opus 35, two Nocturnes Opus 37, and three Mazurkas Opus 41. Chopin wrote only three sonatas and two concertos.

The bulk of his piano works came under headings that were new and strange in his time for a composer of serious purposes. Titles which he chiefly used include Mazurkas, Polonaises, Waltzes, Rondos, Impromptus, Etudes, Nocturnes, Scherzos, Preludes, Fantasias, and Ballades. All these, except the Ballade, which he invented, were established musical forms when Chopin took them over (Leonard 1946).He composed the F Sharp Major Impromptu, Opus 36 when he was quite depressed during this time. He became quite sad and returned to Paris in September 1839. He found an apartment at 5 Rue Tronchet. He was convinced that he had "consumption," with frequent severe coughing and spitting blood. This was consistent with advancing tuberculosis. He led a quiet life from 1840 to 1841 and was not composing much music.

He met Eugene Delacroix in 1840. Delacroix became a good friend and was primarily responsible, through his influence, for Chopin's resurgence of compositions during the next phase of his life. The F Sharp Minor Polonaise, Opus 44 initiated this. Chopin's creativity in 1842 emphasizes again how free he was of patterns of distinct "creative periods" that might have reflected any particular mood or psychological condition. One finds, instead, extraordinary variety, ranging from the powerful patriotic chords of the A-Flat Major Polonaise and the nostalgia for Poland in the Mazurka in C-Sharp Minor, op. 50, no. 3, to the lyricism and poetry of the Scherzo no. 4 (Szulc 1988). "Chopin has written two adorable mazurkas that are worth more than forty novels and say more than the whole literature of the century" (Zamoyski, 1980).

His health continued to deteriorate in early 1844, and it became more difficult to continue his work, given his constant cough, chest pain, and spitting up blood. Dr. Papet, however, made a full examination and found "no signs of illness or damage, but thought him inclined to hypochondria and destined to be perpetually alarmed until he reaches the age of forty and his nerves lose some of their excessive sensitivity" (Zamoyski 1980). George Sand identified a severe, overwhelming depression as she continued to be his main caregiver. Chopin learned that Mendelssohn had died in Leipzig on November 4 at the age of thirty-eight. He was now thirty-six, and of course he knew that Schubert had died at age thirty-one. With the severity of his health problems, he obsessed about an early demise. He could never quite make up his mind about his personal life (Szulc 1998). He became more depressed, and it became more and more difficult for him to work. Music and the devotion of his friends sustained him until the end, particularly after his separation from George Sand. With his separation from Mme. Sand, Chopin's composing ceased, and his deteriorating health accelerated (Leonard 1946).

Chopin was obviously a very shy and quiet man who was an exceedingly gifted creator and performer. He was highly emotional, as expressed in his music composition.

Miss Stirling sent him secretly a gift of twenty-five francs, which sustained him to the end of his life. Chopin's charmed and unhappy life was always concealed from full sight by a mantle of privacy and mystery that was hesitation and uncertainty compounded. He was all things to all people: first as a genius in the world of music, then as friend, a lover, and even as an enemy. Chopin had given so much joy and fulfillment to others through the inspiration and magnificence of his music and taken so little for himself, always the self-deprecating figure.

His closest companion during his final years continued to be Delacroix and they talked about life, art, and music. He moved from Paris to Chaillot for the summer of 1849. Dr. Criveilher diagnosed terminal tuberculosis with pulmonary hemorrhage and leg swelling. Chopin died a great but very sad man who was very much alone (Szulc 1998). He had asked that his body be cut open and his heart sent to Warsaw and that Mozart's *Requiem* be sung at his funeral. He died on October 17, 1849.

CHOPIN AND ACCELERATED COGNITIVE ABILITY

Chopin's life as a genius was driven by an insatiable need to compose music. He needed to express his robust internal emotions and chose the piano as his art form. He needed to express himself as an artist, which was clearly the whole force of his life. He was exceedingly inquisitive and restless from a very early age, which is quite typical for persons of ECA with genius. He longed for the peace of Nohant so that he could compose in quietude. He was quick-tempered and irritable.

Miss Zofia Rozengardt wrote a very revealing letter, which clearly describes his daily struggles with his gift of genius and the discomfort that it usually brings to those so gifted. The personality characteristics that she so able described for Chopin are all quite common in many persons with ECA. She writes that Chopin was "a strange, incomprehensible man! You cannot imagine a person who can be colder and more indifferent to everything around him. There is a strange mixture in his character: vain and proud, loving and yet disinterested and incapable of sacrificing the smallest part of his own will or caprice for all the luxury in the world. He is polite to excess, and yet there is so much irony, so much spite hidden inside it! Woe to the person who allows himself to be taken in. He has as extraordinarily keen eye and will catch the smallest absurdity and mock it wonderfully. He is heavily endowed with wit and common sense, but then he often has wild, unpleasant moments when he is evil and angry, when he breaks chairs and stamps his feet. He can be as petulant as a spoiled child, bullying his pupils and being very cold with his friends."

There were usual days of suffering, physical exhaustion, or quarrels with Madame Sand (Zamoyski 1980). He was a heavily burdened, introspective artist bewildered by his emotions, his art, and his powerful need to express it in music. This unrelenting energy and focus were his life, the power of which was all–consuming and overwhelming at the expense of all other personal endeavors. He was unable to live a normal life, so common in persons of genius. . His social relationships, especially as related to George Sand and other professionals and social relationships, may have had a major impact on his work. But one has the feeling that his core being—his emotional, inner life as expressed in his music compositions—was little affected by all of this other than a need for contact or general support from the real world on an ongoing basis. The major issue, as it is in all creative individuals, is what was the driving continuing force that remains so clear in the absence of any obvious materialistic or monetary return from the immediate world around him—or future rewards for that matter? How important was the world that surrounded him in terms of impacting his composition one way or the other? It seems to me that the creative product and the energy behind it fuels the accomplishment way beyond any obvious external stimulus or reward. He had an inner need to do this, and it may well have occurred anywhere in the world with any kind of social environment around him. What role does his chronic illness play in all of this? What role do the environment and social contacts play in this?

Some will say that Chopin's lasting fame can be attributed to his unhappy existence and to the illness that held him in bondage for most of his adult life and eventually killed him, to the torment he suffered for the Poland that he worshipped, and to the chagrin of an incomplete life because he never achieved the domestic happiness that he yearned for (Murdoch 1935).

Chopin was a very private person who loved solitude. He was quite polite and modest in public but could be quite abusive in letters about fellow composers, pianists, and the French government. What motivated him to continue as a composer, given all this illness and unhappiness? He had major physical and mental illness all his life, coupled with a severe mood disorder: depression. This major illness affected him in many ways not only compromising his time but also significantly affecting his social life and relationships with women. The mood swings may have been important for times of creativity and may have been essential for the depth of his emotion as expressed in his music. It seems that geniuses pay a terrible price for their gift, both physically and emotionally, but they may recognize it as part of what they do and go forward anyway because of the intense need to create at all costs that the internal energy present at birth generates. It is as if they have no choice and must pour their lives into their work to respond to an inner need that goes way beyond their control. (See materials in basic science section related to a unique neurobiological template at birth secondary to emergenesis.)

As is true for most persons of genius, it is not obvious why Chopin would continue to do what he did to the degree that he did it, given the heartache he had to experience. What was his reward for all this? What is responsible for this fierce drive in these individuals? He continued despite his health and being far away from home without solid relationships. There is no apparent gain. It is as if their existed an inner tension that had to be released at all costs and was more important than any other obvious comforts. This may be what genius is all about and may explain the powerful force in these individuals to move forward despite the obvious obstacles.

CHOPIN: FEATURES OF COMMONALITY AND DISSIMILARITY WITH OTHERS WITH ACCELERATED COGNITIVE ABILITY

He had many features in common with other highly creative individuals, including precociousness, extraordinary creativity early, chronic illness with early death (age thirty-nine), emotional insecurity, suspiciousness, clinical manic-depressive illness, reclusive, ability to intently concentrate, endowed with wit, anger, and remarkable perseverance to continue to create despite multiple physical and mental obstacles without any apparent material gain. In general, his profile was typical for this unique group. (See specific personality characteristics Miss Rozengardt outlined, above.)

Perhaps slight variations include a supportive family, emotionally and intellectually, along with a very early sophisticated musical exposure. This is unusual in this group. He was extremely sensitive and quite meek in his demeanor, which is also unusual in others with highly creative personalities.

CHOPIN AND HIS LEGACY

Astolphe Custine captured the essence of the Chopin's legacy when he wrote his musings back to Chopin after his concert February 16, 1848 in Paris arranged by Pleyel. "You have gained in suffering and poetry; the melancholy of your compositions penetrate still deeper into the heart; one feels alone with you in the midst of a crowd; it is no longer a piano, but a soul, and what a soul! Preserve yourself for the sake of your friends; it is a consolation to be able to hear you sometimes; in the hard times that threaten, only art as you feel it will be able to unite men divided by the realities of life; people love each other, people understand each other, in Chopin. You have turned a public into a circle of friends; you are equal to your own genius" (Zamoyski 1980).

In spite his short life of thirty-nine years, which were filled with great discomforts, physically and emotionally, he was able to compose astonishingly beautiful music that originated deep within his soul and left his gift as a legacy of his genius for the ages. He did compose three sonatas and two concertos along classical lines, but the bulk of his work involved polonaises, waltzes, rondos, impromptus, etudes, nocturnes, scherzos, preludes, fantasies, and ballades for the piano, all of which have remained treasures for the musical world. His legacy of passionate, romantic musical compositions remains available, and all of us can, in some deep part of our beings, identify with and cherish. It's as if he captured a divine truth that we all recognize but cannot create for ourselves. He was truly the "tone poet" of the ages.

"It was the best of times; it was the worst of times."
Charles Dickens

"Whatever I have tried to do in life, I have tried with all my heart to do it well; whatever I have devoted myself to, I have devoted myself completely; in great aims and in small I have always thoroughly been in earnest."
Charles Dickens

Charles Dickens CHARLES JOHN HUFFAM DICKENS

INTRODUCTION

Dickens was the "novelist of the age," but why was his intellectual energy directed at writing literature rather than other intellectual pursuits, such as music, art, science, or statesmanship? The spark that leads to extraordinary cognitive ability may well be genetic, and for Dickens, it may possibly lead directly to his grandparents, at least to some extent. "There is no doubt that, in the lives of writers, the shadows of a grandfather or a grandmother can be seen lying across the paths they follow" (Ackroyd 1990). He had a very unhappy, lonely, and neglectful childhood with a cold, fearful mother and a very inattentive father who did not care about him and did not even want him educated. Despite this humble and difficult beginning, Dickens had insatiable curiosity and an inherent gift of observation coupled with a remarkable ability to convey the human condition through written stories. He became the world's foremost novelist of his time. His legacy lives on and continues to inspire writers and provide vivid descriptions of nineteenth-century English society with remarkable real-life portraits. As is typical for most creative persons, his life was colored with certain periods of unhappiness as he struggled with personal societal interactions; his high mental energy, which was difficult to control at times; and his major concern as to social conditions of his day. His story is a great example of genius with ECA at work, as it relates to not only its origins, but also its development over a lifetime and the profound legacy left as a result for future generations.

BIOGRAPHY

He was born Charles John Huffam Dickens on February 7, 1812, near Mile End Terrace at the outskirts of Portsmouth, England to John and Elizabeth Dickens. He had a sister, Fanny. His father was a navy paymaster by trade. His mother's maiden name was Barron, and it may be from her that he received the gift of keen and rapid observation. Her parents were both clerics and makers of musical instruments. Dickens's paternal grandfather was a butler for the Creve family, and his paternal grandmother was known as a great storyteller. The origin of genius may well have skipped a generation as the influence of the grandparents seem much more direct than the contribution of his parents, although his mother was quite inquisitive, and his father may have accounted for some of his grandiloquence, humor, and rhetoric. John Dickens, his father, was negligent and incompetent and ignored young Charles.

In childhood, it was obvious early on that Charles had an intense awareness and inquisitiveness, coupled with a remarkable memory and the ability to retain distinct visual imagery. "He was a voracious reader and would sit with his book in his left hand, holding his wrist with his right hand, and constantly moving it up and down...as if reading for life" (Kaplan 1988). His awareness of his ability to write was recognized early in life as he began to write at age eight. His later novels are full of these retained images from childhood, with much of his later work biographical in nature or at least personally visualized or experienced. It remains unclear why he chose writing at an early age as opposed to other pursuits, although with his magnificent memory, powerful verbal skills in writing and pantomime, and unique visual curiosity, it may not be that much of a mystery.

His mother was his first teacher; however, his father did not want to educate him and was content for him to begin working in the factory, the Warren Blacking Warehouse. The young Dickens already had an audience and sensed the power of words. He enjoyed the praise that his performances elicited (Ackroyd 1990). The following is a direct quote from his novel David Copperfield reflective as an autobiographical revelation of how important omnivorous reading was for him his entire life. "Reading was my only constant comfort. When I think of it, the picture always rises in my mind of a summer evening, the boys at play in the churchyard and I sitting on my bed, reading as if for life."

Books indeed were his comfort and solace for him for his whole life, especially in the atmosphere of a troubled childhood. He was alone and friendless and spent many hours as a child walking by himself. His sister was sent off to the music academy for piano training, his father was in debt, and there was no opportunity for education for him. His father went into debtor's prison at Marshalsea, and Charles worked at the Warren Blacking Warehouse. His novels, *Oliver Twist* and *David Copperfield,* were very much autobiographical as he keenly remembered the vivid details and imagery of his own experiences.

His father enrolled him at Wellington House Academy for two years at ages thirteen and fourteen. He remained high-spirited and continued to pursue his interest in the theater and reading. Because of the continuing debt of his father, his sister, Fanny, had to leave the music academy, and he had to leave Wellington House Academy at age fifteen. He never did attend the University as a consequence. It was clear early on in his life that he had a remarkable visual memory, which was so important for releasing his intellectual energy in written form. He had a very keen mental awareness with a great gift to mimic and create scenes from his childhood.

In 1827, he became a law clerk for Mr. Edward Blackmore at Ellis and Blackmore but became bored and began to educate himself in the art of shorthand, hoping to pursue a career in parliamentary gallery reporting. In 1830, he began to study hard, with voracious reading habits, at the British Museum and developed a romantic relationship with Maria Brednell that lasted about four years. This relationship helped serve as a motivator for a career in journalism.

He became the most powerful writer of his period and infused the whole struggling lower and middle classes into his numerous writings. His relationship with Maria Brednell ended in March 1833 with her rejecting him. His first writing was "A Dinner at Popular Walk," published in the monthly magazine, as he continued to do shorthand as a parliamentary gallery reporter. He became engaged to Catherine Hogarth in May 1835; at the time, he was supplying "sketches of London" for *The Evening Chronicle*, whose editor was George Hogarth, Catherine's father.

His observation and writing skill began to be apparent at this time. He had a keen awareness of the world around him, and one can see in miniature, the formulation of the artist reacting to the events of his life and using them and being used by them (Ackroyd 1990).

His first novel was the *Black Veil*. He remarks about his excitement for the subject matter of which he wrote is typical of a highly energized person with extraordinary cognitive ability responding to a very high internal drive with absolute focus. "I can never write with effect—especially in a serious way—until I have got my steam up, or in other works until I have become so excided with my subject that I cannot leave off...." (Ackroyd 1990). There was a publication of stories and "sketches of Boz" in 1836 and the initiation of the *Pickwick Papers* on February 18, 1936, with Seymour as illustrator. Browne later replaced him.

Dickens was always anxious about money but had no interest in becoming a member of the upper class. Ackroyd believed that Dickens's writing power so well displayed in his numerous novels derived from personal observations laid in his ability to create symbolic narrative rather than in dialogue as stated in a direct quote from his Dickens Biography. "His gift lay in symbolic narrative rather than in dialogue, in creating characters who dwell in language rather than ones who dwell upon the boards" (Ackroyd 1990). At about this time, he developed many new important relationships, including Foster, Maclise, Macready, Ainsworth, Beard, and Mitten. He did not mix with other writers or intellectuals as he preferred to be around persons without the same skill set.

Catherine had problems with severe postpartum depression. He developed a close relationship with her sister, Mary Hogarth, who died at the age of seventeen in May 1837. He described this as the most painful loss of his life. He was twenty-five at the time.

Dickens was alert to every impression and was absorbing as much local detail as he could, knowing all the time that such specific truths would buttress the larger designs that he already had in mind (Ackroyd 1990).

Oliver Twist was created in 1837 and became a monthly serial in *Bentley's Miscellany*. As mentioned earlier, his ability to retain every impression with great detail, which provided specific truths that reinforced his perspectives of the social conditions that continued to revolve about him were reflected in his ability to write wonderfully descriptive novels full of symbolism. He was very quick witted, liked to tell jokes, and developed a real art of story telling and mimicking. He had a fixed routine to his day, with writing in the morning, walking in the afternoons alone or with Foster, supper at five, and then more work or a trip to the theater in the evening. He walked twenty to thirty miles each day as he continued to write (Ackroyd 1990).

"Streams of people apparently without end poured on and on, jostling each other in crowd and hurry forward..." and in this jostling London, Dickens elbows his way forward, identifying within it all his won contrast and contradictions. "Life and Death went hand in hand; wealth and poverty stood side by side; repletion and starvation laid them down together" (Ackroyd, 1990).

The sequence of his books was: *Pickwick Papers,* then *Nicholas Nickleby,* then *Oliver Twist,* then *Barnaby Rudge. Nicholas Nickelby* was completed on September 20, 1839. In 1840, while living in Devonshire Terrace, he lived with his wife, four children, four maidservants, and one manservant. He was always writing at a fever pitch, with up to 2,300 words per day and four chapters in six days. He went to Scotland in 1840 with Catherine, as he wanted to take a year off after completing Barnaby Rudge.

He was haunted by "visions of America," Washington Irving invited him to come to America, and Dickens wanted to see a country that, much like himself, had risen unimpeded from all trials and circumstances. He went to America on December 2, 1842, after completing *Barnaby Rudge*. In America, one "noticed how his large and expressive eyes seemed to be searching every face and reading character with wonderful quickness. Everyone saw that he was able to penetrate each person when he looked at them" (Ackroyd 1990). He had a tremendous reception in America, initially in Boston. He identified with the democratic institutions and the common man but was upset with the lack of adequate copyright laws and compensation in America. He prepared *American Notes,* which was a book of his travel notes, for general circulation. He attended the Unitarian Church, not so much concerned with theology, but with social and moral obligations of faith and public service.

The Life and Adventures of Martin Chuzzlewit was written January 1843. It is interesting that he always began his novels with the name of the character before he developed the character. He continued with a bad temper during this period and became angry with his publisher and family and progressed further into financial debt. The characters he created became all too real for him, and as he feared the magnitude of the ignorance in general and feared that the poor children were not being educated. He had periods of extreme fear, with panic attacks, and worried that he may become poor as he felt that making and retaining money was his defense against eventual poverty, given his experience as a child.

"Dickens was not always the wonderful, devoted friend whom most of his contemporaries seem to remember; there were times when those close to him also became disenchanted with his behavior" (Ackroyd 1990). He had relationship problems with his wife, Catherine, and became infatuated with Christiana Wells, as he had been with

Catherine's sister, Mary Hogarth. He traveled to Italy in 1845, returned to London in June 1845, and became an editor of the *Daily News*. He eventually gave this up to preserve his time for novel writing and turning attention to his passions of societal reform, educational reform, and prison management.

He moved to Paris after completing *Dombey and Sons* in 1848. His next major writing accomplishment was his "autobiography," *David Copperfield*, which appeared in October 1850. It combined a rich mixture of fiction and personal real life experiences for him in the 1820s and 1830s. He began *Bleak House* in 1851, and it was completed in 1853.

He continued to suffer from dizziness, headaches, and severe restlessness with depression on a very frequent basis. He began *Hard Times* in January 1854 and later dedicated it to Thomas Carlyle. The character of his fiction began to change with *Hard Times*, becoming more serious but bringing forward the social problems of the age revolving around the poor, education, and sanitation.

He lived in Paris in 1855, liked public speaking, and began *Little Dorrit*. He separated from Catherine and continued with an agitated depression and with feelings of exhaustion and paranoia. In 1858, he began *A Tale of Two Cities*, after voluminous reading, and by 1859, he was recognized as a man of genius, a writer who had changed the shape of the English novel.

A Tale of Two Cities was completed in October 1859. *Great Expectations* was started in September 1860. His main audience was the low- and middle-class and was not a favorite of the elite. He was not popular with other literary and artistic figures of his day. He continued with complicating medical problems that now included a possible stroke with left side weakness complicated by a facial neuralgia, renal colic, and high blood pressure. He was difficult to be around with lesser literary figures as he was intimidating.

Our Mutual Friend was written in 1865 and was his last complete novel. *The Mystery of Edwin Drood* was begun in 1869. He combined it with multiple public readings, which gave him great comfort. Despite his poor health, he wanted to return to the United States in 1867 with his last public reading in Boston on April 8, 1868.

He remained quite depressed and fearful, especially as it related to the potential lack of income. Back in England, he continued public readings of his material. "Even now, ill, prematurely old, depressed, as he so often announced to himself to be, he was revived by the momentum of his reading scripts and literally forgot himself in the excitement of talking on the very characters he himself had created" (Ackroyd 1990).

He had more small strokes and developed a left visual field defect and a speech anomia. His son, Charles, thought that reading the chapter about Nancy's murder by Bill Sikes from *Oliver Twist, led* to so much emotion for him that this specific reading greatly contributed to his death.

DICKENS AND ACCELERATED COGNITIVE ABILITY

Dickens had many of the common attributes or features of those with ECA as is evident with scrutiny of his life. He was born with a unique curiosity and tremendous visual retention skills. He had a magnificent memory, as his ability to later remember minute details of his early life and incorporate them into his later novels shows.

His power of observation and retention was the source for his genius in writing, especially with character development. He would vividly recall life experiences that he had throughout his life crucial to his later writings incorporating them into his powerful novels full of symbolism. His powerful skills of deep observation, coupled with his complete memory retention, allowed him to approach his writing with full, vivid warehouse of images and ideas.

He had a high degree of inner tension as his high intellectual energy manifested, being hypomanic most of his life. He had some obvious inheritable features from his maternal grandparents, which is unusual but more common with writers than other pursuits of the intellectually gifted. He was constantly reading, even in his youth. He had a keen sense of wit, liked to act on the stage, and could communicate long, detailed stories verbally. He had no formal education after the age of fifteen and never attended college. His passion was writing, acting, and social activism. His schedule was disciplined, with writing in the morning, walking or riding in the afternoon, and attending or participating in plays in the evening.

DICKENS: FEATURES OF COMMONALITY AND DISSIMILARITY WITH OTHERS WITH ACCELERATED COGNITIVE ABILITY

Dickens had multiple features in common with others with ECA, as suggested above. His early, unique, and profound inquisitiveness coupled with his lifelong, keen sense of humor is a characteristic of the intellectually advantaged. He had a high internal energy and was considered hypomanic clinically.

Features of dissimilarity from others with ECA might include a fair degree of family members with similar skills, such as a ken sense of wit and storytelling, although skipping a generation as mentioned. He was socially aware and keenly interested in the plight of the lower and middle classes that he thought the more elite members of society were exploiting.

DICKENS AND HIS LEGACY

He became the greatest English language novelist, with an ability to capture the time in which he lived with historically accurate narrative. But more important, he captured the sense of the culture and human relationships that existed, thus providing subsequent readers with the experience of reliving the age with complete understanding.

Dickens had more than seen these experiences; he had felt them, experienced them, and declared them in his fictions. From a distance, then, he embodied the period from which he sprang, and in the course of his writing career, it was with particular genius that his life itself became an emblem of that period instinctively, almost blindly, to dramatize an age (Ackroyd 1990).

"Art is a human activity having for its purpose the transmission to others of the highest and best feelings to which men have risen."
Leo Nikolaevich Tolstoy

"There is no greatness where there is no simplicity, goodness and truth."
Leo Nikolaevich Tolstoy

"The two most powerful warriors are patience and time."
Leo Nikolaevich Tolstoy

LEV NIKOLAYEVICH TOLSTOY

INTRODUCTION

Born to wealth but rejecting the materialistic world throughout his life, Tolstoy became the most accomplished literary figure of his time and perhaps one of the most powerful minds of all time, particularly with his focus on realistic fiction. He clearly was intellectually advantaged, obvious from a very early date in his life. His philosophy was similar to that of Rousseau, whom he greatly admired who also expressed his hatred for artificiality, pretense, and convention. He was always a very independent thinker, critical of government, most writers, the materialistic society that he experienced in the Russian aristocracy, and the church.

The origins of his literary prowess are not known; however his mother's family was apparently quite intellectual. He was clearly highly intelligent from a very early age but spent most of his life in turmoil, struggling with self-criticism and severe inner tensions as he attempted to better understand life and how it should be led. His religious writings later in life display the severity of his inner uncertainty and his identification with a supreme being. At end of his life, he retreated into solitude and gave up all his materialistic possessions, literally, by leaving home some days prior to his actual death.

His life had many similarities with other persons of extraordinary intelligence but also some important differences, as will be discussed in a later section of this biography.

BIOGRAPHY

Lev Nikolayevich Tolstoy was born into a Russian aristocratic family on September 9, 1828, at Yasanaya Polyana, a Volkonsky family estate, in the province of Tula, near the center of European Russia, south of Moscow. His father, Nikolay, married Princess Marya Volkonsky from a family distinguished for their "intellectual brilliancy" and aristocratic means. There were five children in the family, with three older brothers and one younger sister. His early exposures were typical for the Russian gentry of his day, with leisurely summers spent in the country and winters spent in the large cities. Importantly, he lost his mother at age two and his father at age nine. He obviously did not know his mother, and his father's early death created great feelings of despair for him. He, his sister, and three older brothers were cared for by two great-hearted women, their Aunt Tatiana, of whom Tolstoy said that "she had two virtues: serenity and love." The other woman was their Aunt Alexandra, who was forever serving others. Her favorite occupation was reading the lives of the saints or conversing with pilgrims or the feeble-minded.

Early on, the germs of his future genius were apparent with his great imagination, intense curiosity, and high sensitivity. His brain was always busy, always trying to discover what other people were thinking or had thought. He had precocious powers of memory and observation; an attentive eye, which even in the midst of his sorrow scrutinized the faces about him and the authenticity of their sorrow. His personality profile was quite evident from an early age and he was noted to be quite passionate, jealous, vain, affectionate, and impressionable as a young boy. (Noyes 1918) His brain was in a condition of "perpetual fever," and he suffered from constant self-analysis. He read extensively as a young man, and among his favorite and most influential reads were the Sermon on the Mount, the book of Matthew, Rousseau's *Confessions* and *Emile*, *David Copperfield* by Dickens, and *Eugene Onegin* by Pushkin.

He spoke Russian, French, German, and English but resisted formal education and left the University of Kazan after three years in 1847. He was attracted to gambling and became significantly indebted by age nineteen. He returned to Yasanaya Polyana to seclude himself in the rural setting from 1847 to 1851. He worked as a "yunker" or volunteer officer and later became a member of the Cossacks in 1851. He became and always remained suspicious and critical of all formal institutions, including higher education, the military, government, and later, formal religions. He preferred to be solitary most of his life, except for being with his family.

His first major literary work was *Childhood*, which he wrote at age twenty-three in 1851, and it was very close to an autobiography. He joined the Russian military forces in Bucharest in March of 1854 and served as an officer. He later requested to be sent to the Crimea. Out of this experience, he wrote *Sevastopol*, in which he was highly critical of war and the emotional turmoil and hardship that it created. "You will see war, not in regular, beautiful, and brilliant ranks, with music and the beating of drums, with waving banners and generals on prancing steeds—you will see war in its true expression, in blood, in sufferings, in death" (Tolstoy 1888 and Noyes 1918). He saw war as suffering and death, unrelieved by any touch of brilliancy or grandeur, and he selected just the right detail to convey his impression (Noyes 1918).

This work was quite similar to *The Red Badge of Courage* written by Stephen Crane in the United States a bit later in the nineteenth century, both referring to the apparent insanity of war. The following is from Tales of the Caucasus written by Tolstoy to this point. "Is it impossible, then, for men to live in peace, in this world so full of beauty, under this immeasurable starry sky? How is it they are able, here, to retain their feelings of hostility and vengeance, and the lust of destroying their fellows? All there is of evil in the human heart ought to disappear at the touch of nature, that most immediate expression of the beautiful and the good" (Rolland 1911)). Tolstoy lived through youth in a delirium of vitality and the love of life. He embraced nature all his life.

The Cossacks was Tolstoy's first attempt at a novel. He temporarily joined a literary group in St. Petersburg in 1855, the Contemporaries, but left the group two years later after becoming generally disenchanted with all formal organizations in general. During the next years, he spent time in Paris, developed a stormy relationship with the Russian writer Turgenev, and latter went on to Lucerne Switzerland to visit his aunt Countess Alexandra Tolstoy, who remembered him at that time to be "unaffected, modest, and quite engaging." (Noyes 1918)

After foreign travel, he returned to Yasanaya Polyana in August 1857, where he remained for the next three years until he again left Russia to study public education with his sister, Marya, in multiple European cities. Peasant education remained a keen interest for him, and he devoted a great deal of energy to a peasant school on his estate of his own design. He often found himself at odds with the government and liberal politicians over the methodology and general design of primary education.

The great sin of modern education, according to Tolstoy, was that it was founded on compulsion, and the government was forcing it on people who did not desire it but who desired something quite different (Noyes 1918). His individualism was expressed as contempt for traditional and accepted authority.

In 1863, he moved away from education as his primary focus and began as a writer in earnest, devoting full attention to it with his wife, Countess Tolstoy. She became his biggest supporter and proofread most of his works. He began *War and Peace* in 1863, which many consider the greatest novel ever written. It was written during one of the happiest and most productive periods of Tolstoy's life, and the happiness can be attributed to his very successful marriage early on. The novel traces the history of Russia during the years 1805 to 1812 through five families, all belonging to the higher circles of the Russian aristocracy. The novel is about individuals and is not a historic epic. Common soldiers are the real heroes of the novel. Noyes expands on this recurrent Tolstoy theme of the primary importance of individual direction of life as opposed following the leadership of others referring to this novel about Russia. "Napoleon was defeated by a whole nation of which Kutuzov, the Russian general, was the accidental representative...A man's true mission is to shape his own life; when he attempts to guide external events he becomes ineffective" (Noyes 1918).

After its completion; he studied Schopenhauer, whom he thought was the greatest genius of humankind. Tolstoy's wisdom was absolute submission to the powers that be, absolute refusal to force one's will on a fellow creature, absolute truthfulness, and, above all, universal kindliness and love.

He developed so many concerns about Russia, especially about its formal institutions and rigorous court system, that he at one time seriously considered leaving Russia and moving to London, England.

He began *Anna Karenina* in 1873, and it is a more perfect work, the work of a mind more certain of its artistic creation, richer in experience, a mind for which the world of the heart holds no more secrets. But it lacks the fire of youth, the freshness of enthusiasm, the mighty pinions of *War and Peace* (Rolland 1911).

He remained secluded on his estate after writing arguably the two finest novels ever written in Europe and shunned the literary community in Russia. After writing *Anna Karenina*, he turned away from writing and focused on developing new religious concepts, leading to the composition of *Confession*, which is the story of his own conversion in 1879. This came after a prolonged personal crisis. He thought that the basic unanswered question was what is the true aim of life? "I was not fifty," he said. "I was loved; I had good children, a great estate, fame, health, and moral and physical vigor. I could reap or mow like any peasant; I used to work ten hours at a stretch without fatigue. Suddenly my life came to a standstill. I could breathe, eat, drink, and sleep. But this was not to live. I had no desires left. I knew there was nothing to desire. I could not even wish to know the truth. The truth was that life is a piece of insanity. What permanent meaning can there be to man's existence" (Rolland 1911)?

This pondering of life's ultimate, fundamental question is the root of Tolstoy's entire religious system. The solution for him must necessarily come from faith, which will connect man's finite life with an infinite God.

As mentioned previously, he avoided all formal religions, particularly their institutions, because of moral objections to its practices. He abandoned the church forever in 1878 and sought answers to his basic questions by reading the gospels. He stated that "no man's life vanishes through his corporeal death; it lives on in the memory of him in his influence on other men." (Tolstoy 1884)

Tolstoy's developed system was in thorough accord with the temperamental tendencies that he had shown all through his life. It roots may be defined as individualism, a dislike of civilization, and a Rousseau-like passion for a return to nature, pessimism, asceticism, and love. The perfection of one's inward character is the true aim of man. He saw the chief evil of civilization in the fact that it forces a man to exploit the labor of his fellow men, since without such exploitation, riches and idle ease are impossible. He greatly admired Henry David Thoreau, and their philosophies were very similar. He became a great apostle for international peace, kindness, and universal love. Tolstoy was a literary genius and was able to convey in an art form all the emotion of aristocratic and peasant Russian life. This literary ability is quite similar to that of Charles Dickens's extraordinary ability to capture the life and emotions of common people and to convey them in a similar art form. In him, life and art were one. His artistic work had the value of an autobiography.

Tolstoy left his home, never to return, on November 10, 1910. His friend and physician, Dr. Makovitsky, accompanied him. He left because of a conflict between his faith and his family surroundings, specifically, a clash of his new ideals with those of his wife. He died ten days later on November 20. He remained an individualist who was critical of most societal institutions up to the day of his death, but he had a remarkable ability to not only understand human emotions and interactions but also a powerful talent in his ability to convey these observations in his literary creations.

TOLSTOY AND ACCELERATED COGNITIVE ABILITY

He had a keen and perhaps unique curiosity as a child in the world around him, which is quite common in those individuals of extraordinary cognitive ability. He was sensitive and passionate, engaged himself with great vigor in a variety of social issues early, and read extensively. He decided early on that formal education had many pitfalls for him and all learners and was self-educated as a result. He protested the educational framework of his day in Russia vigorously, which may reflect his individualism and confident ability to better educate himself than by any formal educational process. He remained intellectually engaged his entire life, with expansive reading and compositions of his literary work. His uncertainty as to the meaning of life and his particular purpose continued to provide him with great intellectual stimulation in the later years of his life.

TOLSTOY: FEATURES OF COMMONALITY AND DISSIMILARITY WITH OTHERS WITH ACCELERATED COGNITIVE ABILITY

He had a typical, very early keen curiosity coupled with a voracious appetite for reading that is most common in these individuals. His sense of independence and self-confidence were quite apparent, and he was always vigorously engaged in the issues of the day. He was reclusive, which also is quite common, and was quite intolerant of others in the intellectual arena that he deemed followers and not independent thinkers. There are no clear areas of dissimilarity with others with this extraordinary cognitive ability that I can discover, other than perhaps his very long life and his apparent emotional control, except for near the end of his life.

TOLSTOY AND HIS LEGACY

He was Russia's greatest writer and was most able to convey the typical Russian personality of either the aristocratic or the peasant class. Malcolm Cowley, in his introduction for *Anna Karenina,* best specifically described the genius of Tolstoy by stating that "genius is energy—mental energy first of all, physical, emotional, and sexual energy, along with vision often involving this gift of patterns where others see nothing but a chance collection of objects. Genius is a memory for essential details; genius is the transcendent for brooding over a subject until it reveals its full potentialities. Genius is also belief in oneself and the importance of one's mission, without which the energy is dissipated in hesitations and inner conflict." (Tolstoy 1960). He was a novelist par excellence with *War and Peace* considered to this day the best novel ever written. Through his writings, today we understand nineteenth-century Russian culture. He was a master of realistic fiction and was able to convey these intimate observations to the general reader through his novels. "It is true that each loved him for different reasons, for each discovered in him himself; but this love was a love that opened the door to a revelation of life; to the wide world itself. On every side—in our families, in our country homes—this mighty voice, which spoke from the confines of Europe, awakened the same emotions, unexpected as they often were" (Rolland 1911). His writings record all our passions in a way that we can all recognize but would be unable to create. As true for many great artists, his life and art were one.

"Nothing in life is to be feared, it is only to be understood. Now is the time to understand more, so that we may fear less."
Marie Curie

"All my life through, the new sights of Nature made me rejoice like a child."
Marie Curie

Marie Curie MARYA SALOMEE SKLODOWSKA (MARIE OR MADAME CURIE)

INTRODUCTION

Madame Curie, a late nineteenth-century immigrant from Warsaw, Poland, to France, accomplished remarkable advancements in science despite multiple barriers, a testament to her remarkable intellect combined with a very hard work ethic. She was a genius; her very creative pursuit of a greater understanding of the mysteries of nature through scientific research led to the discovery of two new elements, radium and polonium, and a better understanding of radioactivity, which continues to provide significant new and useful contributions to society.

Marie Curie overcame a very difficult childhood with poverty and limited access to schooling, requiring primarily self-education, during a time of extreme social and political unrest in Poland. However, through persistence she realized her dream of a higher education in France so she could follow her father's devotion to science. She was the first woman to be admitted to the prestigious Sorbonne in Paris, where she pursued physics and mathematics and which evolved into her career of mineral research and thus her scientific achievements. She later received the first Nobel Prize given to a woman, in 1903 in physics, and was awarded another Nobel Prize in 1911, in chemistry. She continued her research and helped found two scientific institutes devoted to radiation research, named in her and her husband's honor.

What is even more remarkable is that she accomplished all of this in a very difficult environment where, given the degree of gender bias then present, women were not thought of as equals—particularly in scientific research. She left a profound legacy of a greater understanding of radio-activity and its implications for medical therapy, the identification of two new elements, and the accomplished scientific methodology she used to do her research.

GENERAL BIOGRAPHY

Great lives in science are all about passion and curiosity, and Madame Curie was one of the greatest scientists of the twentieth century. She was born on November 7, 1867, in Warsaw, in what was then called the Kingdom of Poland and was under the rule of the Russian empire. She had four older siblings, and her parents, Wladyslaw and Bronislava Sklodowska, were teachers. Wladyslaw was devoted to mathematics and physics as well as scientific research. They had lost their property and fortunes through patriotic involvement in Polish national uprisings that attempted to restore Poland's independence from Russia. The poverty and the oppressive rule of the Russian empire made life difficult for the parents and greatly restricted the education and activities available for their children.

Marie was twelve years old when her mother died from tuberculosis, and she lost one of her older sisters, Zofia, two years later to typhus. After these personal tragedies, she developed a major depression disorder, and feelings of desolation and isolation became a recurrent problem for her the rest of her life. She called these depressive episodes "fatigue" or "exhaustion." Her father was an atheist, and Marie became an agnostic after the deaths of her mother and older sister.

Marie's family was composed of people who loved learning and became priests, doctors, teachers, and musicians. However, Russian officials believed that women should never enter public life or politics, or indeed hold any other influential position in a "man's world." While all of the Sklodowska children were intellectually bright, Marie was the most brilliant. At a very young age, she demonstrated a powerful memory through her ability to write down poems that she had only heard. After graduating from the government gymnasium, she became a governess to help finance her older sister's medical school education, planning to later emigrate to France and study physics and mathematics at the Sorbonne. She continued to have periods of severe depression and had a very unhappy love affair with a man who was not allowed to marry her because she was "under his station in life." (Goldsmith 2005)

When Marie arrived in Paris, she first lived with her sister Bronya, but later moved to the Latin Quarter and lived in a total of four garret rooms during her two-plus years attending school as the first woman admitted the Sorbonne, the prestigious University of Paris. These were very stark living conditions, but finally she was living her dream of studying science, along with having a semblance of liberty and independence. She was unconcerned or unaware that she belonged to the "weaker sex." She graduated first in her class and began to work for her professor, Gabriel Lippmann, focusing on magnetic properties of various steels. She then met Pierre Curie, who invented a number of delicate instruments, including the key electrometer that helped with her metal studies. She did return to Warsaw in the summer of 1894 to visit family, as well as to apply for a position in science at Krakow University, which she was denied because she was a woman. Upon her return to Paris in 1895, she and Pierre were married. Later, they both shared the 1903 Nobel Prize in physics with Henri Becquerel.

She remained very unsocial and continued to have a series of depressive episodes. She and her husband had financial problems, so she became certified as a teacher and taught on the side while she pursued the magnetic properties of steel. She received the Gegner Prize from the French Academy of Sciences for her work—mainly for the precision of her laboratory methodology. She had her first child, Irene, on September 12, 1897, which created significant scheduling issues for her lab work and extracurricular teaching demands, leading to more serious episodes of depression. She refused to go to the sanatorium, as the doctors advised, because she would not leave her work or her family. Fortunately, after Pierre's mother died, his father came to live with them to assist with child duties.

Marie Curie measured many mineral elements with an insatiable curiosity and extreme thoroughness, which involved a very tedious research methodology. By March 1898 she had established beyond doubt that several minerals gave off more energetic rays than pure uranium alone per unit-weight. Later that year she wrote a paper on her work in measuring elements' radioactivity. She found that pitchblende ore had two elements that were radioactive, one which behaved chemically like bismuth and the other like barium. She then used the technique of fractionation to tease out and isolate the presumed separate elements and soon found that these materials were 17 times more radioactive than uranium—and subsequently, 150 times, then 300 times, and eventually 400 times more radioactive than pure uranium. She thought this represented a "new element," which she called polonium, and then confirmed it with spectroscopy. She used the "Becquerel rays" to study these different materials and their degree of radioactivity. In December 1898, she found an element that produced radioactivity 900 times that of pure uranium and called this element radium.

She discovered both polonium and radium in the same year and then set out to isolate them, which the chemists demanded before they would accept them as new elements. This became a very difficult task and would require limitless vision, skill, persistence, and the dedication of an individual who was passionate about her work. She suffered years of backbreaking toil without money and without assistance, enduring the scorn of fellow scientists, to isolate pure radium salt and eventually apply her discovery toward curing cancer. This required a very laborious procedure, distilling tons of pitchblende. The process took about three years and involved eight tons of pitchblende residue, four hundred tons of rinsing water, and thousands of chemical treatments and distillations. Her goal was to isolate pure radium salt and measure its mass. She eventually isolated one-fiftieth of a teaspoon of material, and it became labeled element number 88. She and Pierre presented their findings at the Paris Exposition Universelle of 1900. Ernest Rutherford thought the radioactivity originated from within the atom, but the Curies thought the activity came from outside of the atom.

Pierre Curie and Henri Becqerel were nominated for a Nobel Prize in physics for 1903, but Marie Curie was not mentioned, even though her peers recognized that she was the central person in the work. Pierre said he would not accept the prize if awarded unless Marie was an equal participant in the award. The Curies received the 1903 prize on December 10, 1903, but did not make the trip to Sweden because of Marie's poor health, "probable TBC and Rheumatism" (Goldsmith 2005) aggravating depression. She became the first woman ever to receive the Nobel Prize and was the only woman prize-winner for thirty-two more years, with the exception of her second Nobel Prize, in chemistry in 1911, until her daughter Irene Joilot-Curie won the Nobel Prize in chemistry in 1935.

Marie prided herself on intellectual achievement, not on material possessions, and thus did not pursue financial gains from her important discoveries; but she and her husband became involved in the possible therapeutic medical implications of radio isotopes. In April 1905, Pierre gave a talk to the Nobel Prize committee and gave full credit to Marie for discovering radioactive substances, especially radium and polonium.

Both Pierre and Marie had deteriorating health but remained unaware of the radioactive exposure they were dealing with and its secondary health effects. A hundred years later, Marie's clothes were found to still be radioactive. The Curies' second daughter, Eve, was born on December 6, 1905. Pierre's health continued to deteriorate from the radiation exposure, and he was killed in April 1906 in a carriage–pedestrian accident walking across the Rue Dauphine in Paris.

Marie became very depressed after the loss of Pierre and led a lonely existence. There would "never be joy" (Goldsmith 2005) for her in the future, according to her daughter Eve, and she became almost noncommunicative. She did assume the chair in physics that Pierre held at the Sorbonne, becoming the first woman to ever become a full professor at the university. The laboratory was her "safe harbor." She did not relate well to her daughters, especially the younger Eve. Her father-in-law continued to be the mainstay for the daughters while she continued to work. But with his death in 1911, she became very depressed again.

Marie continued to pursue the use of radium for medical treatments and industrial uses. With the advent of the First World War, she transferred all the radium then available to Bordeaux, where the French government had relocated, and then devised mobile X-Ray units to use on the front lines to more adequately diagnose traumatic injuries. She was separated from her daughters for two years and had l'Arecouest, a Polish governess, take care of them. Irene entered the Sorbonne to study mathematics and chemistry, as well as learning to become a nurse.

In 1921, Marie came to America, where she received some funds to continue her scientific work and met with President Harding. At that time, radium and radioactivity were heralded as the possible cure for cancer, though Marie tried to dispel that idea. Radium was later replaced by cobalt-60 in the mid-1950s as the agent of choice for radiotherapy for cancer.

Marie died of aplastic anemia on July 4, 1934, most likely secondary to heavy radiation exposure over the years from her laboratory work. Her daughter died from leukemia at age fifty-nine in 1956, likely caused by exposure to radiation as well. Sixty years after her death, in 1995, Marie's remains, along with Pierre's, were moved to the Pantheon in Paris in tribute to their magnificent contributions. Marie remains the only women to have achieved such an honor.

MADAME CURIE AND EXTRAORDINARY COGNITIVE ABILITY

Madame Curie had many features consistent with ECA from a very early age. She was a very bright young girl and did well in school, often under very trying circumstances, including the severe gender bias at the time that kept Polish women from obtaining a good education, further magnified by the Russian rule that greatly discouraged young Polish students in general from obtaining a meaningful education. Very early, Marie was attracted to the study of mathematics and physics, which were also the interests of her father, a teacher and scientific researcher. She was primarily self-taught in science, with the help of her father, and became the first woman admitted to the Sorbonne in Paris for graduate-degree work. A combination of very hard work, extreme focus, and a very skilled research technique led her to receive the first Nobel Prize granted to a woman, in 1903 in physics, followed by a second Nobel Prize, in 1911 in chemistry.

MADAME CURIE: FEATURES OF COMMONALITY AND DISSIMILARITY COMPARED WITH OTHER CREATIVE INDIVIDUALS

Marie Curie had some features in common with other creative individuals, including a powerful ability to focus on a passion, coupled with accelerated cognitive ability evident very early in her life. She liked to work alone, and at one time in her life, she became solely devoted to her work at the expense of all social relationships, until she met her future husband, Pierre. She was always a hard worker and sacrificed time with her family for her passion, which continued throughout her life. She had significant emotional issues, with recurrent severe depression, which also is quite common in this group of individuals. She accomplished great strides in science and left a remarkable legacy from her work and her persistence of effort, despite the severe gender bias she had to deal with.

Features of dissimilarity primarily include a strong family influence on her career, with her father as the primary source of inspiration to pursue a career in science. In addition, her own daughter basically followed in Marie's footsteps to pursue a career in physics and chemistry and eventually took over her position at the Curie Institute in Paris. It is quite unusual for children of individuals with clear ECA to succeed at the same career, which her daughter Irene did, reaching the heights Marie had achieved when Irene became only the second woman in history to get a Nobel Prize, in 1935. Marie's life also was clearly different from others reported in this work because of the severity of the gender bias she had to navigate around to accomplish all that she did.

MADAME CURIE: LEGACY

Madame Curie was a scientist of the first order, discovering two new elements with her sophisticated experimental methodology and helping initiate the field of radiation for medical therapy. She was a patriot during World War I with her front-line medical technology activities, her constant quest for peace, and first and foremost her leading by example the movement for the equality of women against the tremendous odds present at the time. She obviously had exceptional cognitive ability with creativity, accounting for all of her many accomplishments that have had a very positive impact on her own and subsequent generations.

"Your time is limited, so don't waste it living someone else's life. Don't be trapped by dogma – which is living with the results of other people's thinking. Don't let the noise of others' opinion drown out your own inner voice. And most important, have the courage to follow your heart and intuition. They somehow already know what you truly want to become. Everything else is secondary."
Steve Paul Jobs

"Deciding what not to do is as important as deciding what to do."
Steve Paul Jobs

STEVE PAUL JOBS

INTRODUCTION

Steve Paul Jobs possessed many intellectual and behavioral features of an individual with accelerated cognition or clinically designated genius and may represent a contemporary example of the ECA phenomenon. He certainly meets the criteria for the clinical definition of genius, with accelerated intellectual functioning and clear creativity beyond the ordinary, leading to novel, useful advances for peers and future generations. In addition, he had many of the characteristics usually apparent in others with clear extraordinary cognitive ability, such as high intellectual energy, intense focus in all his pursuits, high intellectual powers, independence, restlessness with unhappiness, great self-imposed sacrifice, daily heavy burdens, and very little patience for the day-to-day mundane.

He obtained eminence by leaving an impressive body of highly creative and useful contributions for peers and posterity alike. In his recent excellent biography of Jobs, Isaacson quotes Jobs in his own words as to his clear statement relating his lifelong core belief and strategies: "The creativity that can occur when a feel for both the humanities and the sciences combine in one strong personality was the topic that most interested me in my biographies of Franklin and Einstein, and I believe that it well be a key to creating innovative economies in the twenty-first century" (2011).

I think that he lived the passion of combining his interests in the humanities along with the sciences. He was able to imagine and create technical advancements—or at least lead others to create technical advancements—that he knew we all would want to utilize before we knew what they were. He was a hard worker with a vision and developed digital products beyond his peers' wildest expectations in relatively few years.

Biography

Joanne Schieble and Abdulfattah "John" Jandali, Job's biological parents, developed a relationship while students at the University of Wisconsin. Joanne's father was very much opposed to his daughter marrying a teaching assistant from Syria, so when Joanne became pregnant, she decided to go to San Francisco to have the baby and allow the baby to be adopted. The baby, Steven Paul Jobs, was born on February 24, 1955, and was adopted by Paul and Clara Jobs. It was made clear to the adopting parents that both Schieble and Jandali insisted that the child be fully educated, including a college education.

Steve Jobs knew from an early age that he was adopted and was repeatedly told by his adopted parents that they specifically picked him out, so he was very special. Jobs, however, continued to talk about being abandoned and the pain that it caused him throughout his life, and he remained very angry about it. He became quite independent and hostile, especially in relationships with others, personally and with business colleagues. He had a great difficulty controlling himself with reflex cruelty when dealing with others, which may go back to his being abandoned at birth. The real underlying problem was the theme of abandonment throughout his life, although he loved his adopted parents and always spoke very highly of them.

Paul and Clara Jobs moved to Mountain View, California, where young Steve remembered how impressed he was with the craftsmanship of his adopted father, Paul, as he could design and build anything. He fondly remembered Paul as calm and gentle and said that he served as an important role model, especially with his interests and mechanical ability.

Importantly, Hewlett and Packard started the Hewlett and Packard Co. in Palo Alto, California, in 1938. They were instrumental in building transistors using silicon rather than the more expensive germanium, which would later prove crucial to the development of affordable circuit boards for building computers. Steve joined the Hewlett-Packard Explorers Club where HP engineers would discuss light-emitting diodes with students. This was also the venue for his introduction to his first desktop computer.

Jobs learned to read before attending elementary school and was generally bored during his early years in school, which he claimed may have had a lot to do with some behavioral issues he developed at that time. He encountered authoritarian pushback that he claimed came close to almost beating any curiosity out of him. He began using street drugs as a sophomore in high school, along with exploring the mental effects of sleep deprivation. He became quite rebellious and aloof but became friends with Stephen Wozniak in McCollum's electronic class in high school. Wozniak was very knowledgeable about electronics, and it was a relationship that would later prove to be the core ingredient that led developing of Job's visionary ideas and the initiation of the Apple Company.

In 1971, Jobs and Wozniak developed the "blue box," which was a device that could make calls through AT&T with a tone that would allow long-distance phoning without extra charges by creating a 2600-hertz sound that fooled the recognition system. The working relationship between Jobs and Wozniak is quite interesting with a resulting highly successful partnership composed of two very different individuals in terms of their personalities, intellectual focus and overall motivations. Isaacson describes this very well with the following quote. "Wozniak was a gentle wizard coming up with a neat invention that he would have been happy just to file away, but Jobs would figure out how to make it user-friendly, put it together in a package, market it, and make some money" (Isaacson 2011).

Jobs was passive–aggressive about going to college and threatened to go to New York if he could not go to Reed College in Oregon instead of Cal or Stanford. He thought that Reed College, a liberal arts school, would be much more interesting, and his parents consented. Timothy Leary, the guru of psychedelic enlightenment, was at Reed at the time and was suggesting to the students that they should "turn on, tune in, and drop out." He became engaged with Eastern spirituality and Zen Buddhism and began to realize that, for him, intuitive understanding and consciousness was more significant than abstract thinking and intellectual logical analysis.

He thought taking LSD was a profound experience for him and that it was most important to his life. He became a vegetarian and embraced extreme diets including purges, fasts, or eating only one or two foods for a week at a time. He left Reed College after eighteen months to return home to Mountain View and applied for a job at Atari, the video game manufacture. He worked the night shift and was hypercritical of the abilities of the other employees. He then went to India for a serious "search of himself." He was seeking enlightenment through ascetic experience, deprivation, and simplicity. Steve Jobs embodied fusion of flower power and processor power, enlightenment and technology, as he meditated in the mornings, audited physics classes at

Stanford, worked nights at Atari, and dreamed of starting his own business. Isaacson expresses his unifying concept as to the young culture surrounding Jobs at the time with the following quote - "The people who invented the twenty-first century were pot smoking, sandal wearing hippies from the west coast like Steve, because they saw differently" (Isaacson 2011).

Steve and his high school friend, Steven Wozniak, attended the local Homebrew Computer Club, which was primarily composed of counterculture individuals who were passionate about technology. After one of these club meetings, Wazniak claimed that the whole idea of a personal computer came to him, and he sketched out on paper what would later become known as the Apple I. Wozniak wanted to build computers inexpensively and then give them away free, but Steve Jobs wanted to find a way to make money. Apple was born in June 1975 with an engineer from Atari drawing up the circuit boards, which led to printing fifty units. Jobs visited the All One Farm about this time as he was on a fruit diet at the time, and he and Wazniak decided to call the new company Apple.

The Jobs home in Los Altos became the assembly point for the fifty Apple I computer boards, and they were delivered to a local retail outlet called the Byte Shop. This had to be accomplished in thirty days when the payment for the parts became due.

Beyond these first fifty circuit boards, Steve Jobs wanted to build a "fully packaged computer with seamless integration," which required capital to initiate. On January 3, 1977, the Apple Corporation was formed by buying out the old partnership of Jobs and Wazniak that had been established only nine months previous. Mike Markkula was a major contributor at this time, with a $250,000 line of credit for which he received a one-third interest in the new corporation. He remained with the company for several decades in an administrative capacity. Markkula recruited Mike Scott to become the president of the corporation very early on.

Apple II was launched in April 1977 with twelve employees. At this point, they needed much more space and moved out of the garage to a rented office on Stevens Creek Boulevard in Cupertino, California. At this time, Jobs became increasingly tyrannical and sharp in his criticism of others, as he was obsessed with a passion for the product and a passion for product perfection, according to Mike Markkula. The Apple II was sold in various models for the next sixteen years with six million sold. "Wozniak deserves the historic credit for the design of its awe-inspiring circuit board and related operating software, which was one of the era's great feats of solo invention. But Jobs was the one who integrated Wozniak's boards into a friendly package, from the power supply to the sleek case. He also created the company that sprang up around Wozniak's machines" (Isaacson 2011). This again is a beautiful description of a highly successful partnership outcome composed of two completely different individuals perhaps a further reflection to the depth of Jobs visionary skill in combination with his able powers to accomplish that vision.

A separate venture with the Lisa computer was less successful. Relationship issues were surfacing among Jobs, Mike Scott, and Mike Markkula, primarily because of Jobs's disruptive behavior, resulting in Jobs losing control of the Lisa project.

"When Mike Markkula joined Jobs and Wozniak to turn their fledgling partnership into the Apple Computer Co. in January 1977, they valued it at $5,309. Less than four years later, they decided it was time to take it public. It would become the most oversubscribed initial public offering since that of Ford Motors in 1956. By the end of December 1980, Apple would be valued art $1.79 billion, and in the process, it would make three hundred people millionaires. At age twenty-five after the IPO, Jobs was worth $256 million" (Isaacson 2011). In 1981, the Macintosh was born. The idea was to sell it for about a thousand dollars. It would be a very simple appliance with screen, keyboard, and computer in one unit. Jobs took over the design and marketing from Raskin, who was the origina-tor of the idea of a simple, inexpensive computer, as he did not want to risk product quality for producing it inexpensively. Rebelliousness and willfulness were ingrained in his character; he thought he was special, like Einstein and Gandhi. This drove his passion for making a great product and not just a profitable one. He combined his artistic talent with his pas-sion for product perfection followed by a keen sense of appropriate and timely marketing expertise.

He thought that he could build a much better product than the IBM PC, and many think that he indeed achieved this. The launch for the Macintosh was January 24, 1984, and he and John Sculley, who was then the president of Apple, disagreed on pricing. The machine was introduced with great fanfare, and the market and press eagerly embraced it. About this time, Jobs became aware that Bill Gates and Microsoft were developing a new operating system called Windows for the IBM PC. The news greatly disap-pointed him, as he thought that he and Gates were going to work together going forward at this time. In May 1984, the relationship issues between Sculley and Jobs continued to deteriorate. The Apple corporate board was forced to decide between Jobs and Sculley, and they chose Sculley.

On leaving Apple, Jobs attempted to build a computer for the education industry, the NEXT, but this was a flop and was abandoned in 1989. During this time, he also developed a partnership with Disney called Pixar, which became quite successful. His friend, Larry Ellison of Oracle, and he decided that he needed to get back into Apple, and they eventually sold NEXT to Apple, which was his reentry point into the corporation that he founded. "In returning to Apple, Jobs would show that even people over forty could be great innovators. Having transformed personal computers in his twenties, he would now help to do the same for music players, the recording industry's business model, mobile phones, apps, and tablet computers" (Isaacson 2011). He returned to Apple in December 1996 after an absence of eleven years as a part-time advisor. He would quietly wrestle power from Amelio, who was in charge at the time.

Apple lost $1.04 billion by September 1997, but after the fiscal year of 1998, Apple would turn a $309 million profit. Jobs became CEO in September 1997 and became a partner with Jonathan Ive, who proved very instrumental in future product development by Apple over the next decade. He then developed the "Digital Hub Concept" in concert with the development of the Macintosh computer, which would later lead to the development of the iPhone, iPod, iTunes, and iPad.

His health became an issue when he was diagnosed with pancreatic cancer at Stanford Medical Center and was told that he would need surgical removal. Initially, he refused but later submitted to the surgery in July 2004 when he had a modified Wipple procedure. He thought that his health first became an issue when he was in charge of Pixar and Apple back in 1997, when he was under extreme pressure to run both organizations. By 2008, it became clear that his cancer was spreading despite genome sequenced targeted therapy, leading to a liver transplant in Memphis, Tennessee, in March 2009.

At this time however, he was found to have metastatic lesions to the peritoneum. He wanted to live long enough to see his son graduate from high school in June 2012. Bill Gates made a personal visit to Jobs in the spring of 2011 and spent three hours with him. By July 2011, his cancer had spread to his bones and other parts of his body, and doctors were having trouble finding targeted drugs that would help slow the acceleration of the tumor growth. He officially resigned from Apple at the annual board meeting in August 2011.

JOBS AND ACCELERATED COGNITIVE ABILITY

In his very brief life, Jobs had multiple features that seem to characterize individuals with extraordinary cognitive ability. These included an early rebellious nature in childhood combined with a keen curiosity and a voracious appetite for reading. He was clearly able to think on many simultaneous levels and never ruled out anything as impossible. He became aware of his special intellectual gifts very early on, which was embarrassing for him in thinking that he might be smarter than his adoptive father, whom he greatly admired. He had very little patience with others and was generally moody, angry, and in general, colleagues and family considered him difficult to deal with. He was not the technical innovator that some might have thought but was very creative in terms of design, product perfection, marketing, and especially a vision for the future far ahead of his immediate peers. His passion for developing fully integrated, revolutionary, and simple products to fulfill his vision of the future market needs combined with his marketing abilities certainly describes his creative genius. He introduced significant advancements to society.

JOBS: FEATURES OF COMMONALITY AND DISSIMILARITY WITH OTHERS WITH ACCELERATED COGNITIVE ABILITY

His life displayed many features in common with others in history with clear ECA, such as ability to sharply focus, very high cognitive energy with unique curiosity early in life, emotional intensity, little patience with family and colleagues, numerous health issues and shortened life, boredom with school with little impact from formal education, problems with the mundane, no apparent direct genetic links to his past family with exaggerated cognitive ability, great work ethic, independence, extraordinary memory, and marked difficulty dealing with peers.

Features of dissimilarity with others with ECA included a lack of a sense of wit. Jobs was not clearly precocious and was not an artist, composer, writer, or political leader, as many others were. He certainly had a visionary gift and the ability to not only develop the key products but also the skill to market them in ways that were very appealing.

JOBS AND HIS LEGACY

Walter Isaacson best describes Jobs legacy by stating that his passions, perfectionism, demons, desires, artistry, devilry, and obsession for control were integrally connected to his approach to business and the products that resulted (2011). The saga of Steve Jobs is the Silicon Valley creation myth writ large: launching a startup in his parents' garage and building it into the world's most valuable company. He didn't invent many things outright, but he was a master at putting together ideas, art, and technology in ways that invented the future.

DISCUSSION OF BIOGRAPHY SECTION

Highly creative people harbor an impressive array of intellectual, cultural, and aesthetic interests. They're widely open to novel, complex, and ambiguous stimuli; capable of de-focused attention; flexible; introverted; and are generally very independent (Goldberg 2004).

The origin of extraordinary cognitive ability in any individual is obviously quite complex and multi-factorial involving not only the basic pattern of the biological anatomical, chemical, and physiological framework as determined genetically and modified by early development but also the myriad of varied environmental influences of different intensities. All these factors in combination are responsible for the ultimate configuration of the cognitive process and the all–important, definitive focus of the intellectual drive itself. A similar electrophysiological and anatomical template determined in large part genetically, which is the basic underlying core ingredient common to all these gifted individuals resulting in very early similar intellectual and emotional challenges may exist. The diversity of outcomes with this originating commonality obviously exists secondary to the wide variety of very important and varied environmental influences unique to each individual. These key societal and environmental factors need to be further understood as to not only the specific influences of importance but also the crucial factor of timing.

These brief biographies compare common multiple attributes and dissimilarities important to the individual outcomes of this "intellectual gift." All of them have likely occurred from a fundamentally similar neurobiological framework or biological template.

Many areas of commonality are identified with detailed biographic studies of these unique individuals. There clearly is a markedly accelerated cerebral energy with heightened velocity of thought evident shortly after birth and associated with a keen sense of inquisitiveness or curiosity. The intensity of this heightened velocity of thought creates a significant and burdensome sense of inner tension, which persists for their entire lives and thus precipitates very early on psychological adaptive measures in attempt to reduce the intensity of the inner tension. Their unique ability to sharply focus their thinking with an almost complete unawareness of surrounding influences is a very common attribute in this group and accounts for the rather routine absence of "social skills."

Some of these resultant psychological adaptations are more successful than others are, and in some, they may lead to clinically identifiable psychopathological states. This typical and common attribute can often be identified very early in life. Most of these individuals have an associated incredible retention capability, become quite impatient, demonstrate a high degree of intolerance to educators and peers alike, and are frequently identified with significant behavioral problems if not clinically evident psychopathological disorders. Most of these individuals proceed with "unhappy" lives and significant adaptation problems, with a majority displaying periods of significant depression if not clinical bipolar manic-depressive disorders. Extraordinary cognitive ability associated with the accelerated velocity of thinking that is the core ingredient in this unique population predictably has intense inner tension stimulated by the high cerebral energy. This may account for the drive to be creative as a relief for some of the uncomfortable inner tension. This pure creative energy generated perhaps as a pressure relief measure may well account for the continued inspiration and enthusiasm so apparent for most, despite the absence of any obvious material gain in their lives other than the pure joy of being intellectually creative.

The need for solitude seems universal in these individuals, and suicide is more common that one would predict from the general population. It may be reflective of the intensity of inner tension created by this gift. Early formal general education attempts seem of little value if not outright rejected by most, but societal and important environmental exposures seem to have a high correlation to the overall direction or focus and intensity of the intellectual pursuit.

Common characteristic traits include self-confidence, high degree of alertness, intense curiosity, unconventional work and societal behavior, obsessive personalities, hard working, and intolerant with a high degree of impatience. Most require some degree of societal support, but this is relatively limited in nature. This high cerebral energy leads to massive physical and emotional strain for the individual, which may account for the commonality of early medical problems in their lives if not the rather common phenomena of early deaths when compared to the population at large. Does the higher frequency of medical issues, if not chronic illness itself, occur from the intensity of the inner tension, and what role does it play on the creative process? Does it become an important ingredient to the creative drive, or does it remain a primary deterrent or obstacle to the overall success of the endeavor especially in terms of time?

Major early developing chronic illnesses seem common, along with bipolar disorders. Both may be important in understanding not only the creative nature but also the intense energy that seems to be part of the process. One of the more interesting traits for these individuals is the incredible mismatch obvious between outward behavior and societal interactions on a daily basis compared to the depth and delicacy of the internally generated artistic creativity.

What crucial determinates account for the specific focus and associated intensity of the cerebral energy? This becomes a very important consideration in attempting to better understand why some individual pursuits with the presumed same basic biological elements in place direct their energies in destructive directions, as illustrated in the lives of individuals like Joseph Stalin and Adolph Hitler if one accepts they had similar features of ECA. What allows for a Dickens, Tolstoy, or Chopin as opposed to a Napoleon or Hitler? There may be very subtle or perhaps not-so-subtle significant environmental influences at a very early age accounting for this striking variation in outcome with a presumed similar basic biological template at birth.

This then leads to the speculation about perhaps a very large subset of individuals who are neither obviously creative in a positive or negative direction but fall in between and may account for behavioral disorders of significance not easily attributable to this same basic biological template or framework of accelerated cognitive ability. This, of course, would greatly enhance the importance of understanding those environmental factors so crucial in molding the direction and intensity of this greatly heightened cerebral processing very early in life, both in terms of its nature and timing.

One could speculate as to the specific determinates involved at this point, but a clear understanding must await specific detailed psychological studies of important relationship issues very early in development. They may have some commonality but may also have unique features for each individual. Are there more common recurrent themes that might explain presumed behavioral disorders in those individuals who never focus strongly in either a positive or a negative direction? The answer here needs to await comprehensive psychological population studies of various behavioral disorders that display features of extraordinary cognitive functioning.

On balance, one would hope that the positively directed intellectual pursuits with this same biological framework far outweigh those pursuits in a negative or destructive direction. The incredible legacies of some of these individuals for future generations as pointed out in the biographies above are obvious and remain as extraordinary gifts for generations to come. Perhaps the larger issue in terms of legacies may be to better understand the population subset with significant behavioral disorders as part of this ECA phenomenon, with attempts to recognize early and thus positively modify the psychological elements involved.

The basic biological process, coupled with the environmental influences so important as to its intensity, direction, and ultimate legacy, remains a fascination. A basic study of unique human cerebration, individual or population outcomes that environmental exposures individually modified, and acknowledging the respective legacies reveal they are either marvelous gifts or massive destructive forces.

BIOGRAPHY SUMMARY

What did Beethoven, Schubert, and Dickens have in common with many other persons of genius? They had unusual intellectual energy from infancy, high thought velocity, and incredible retention capacity clearly well beyond the average from a very early age. These traits are clear from biographic analysis. All had a very high driving force, perhaps secondary to the velocity of thought and retention ability, directing vigorous pursuit well beyond the norm. All had very unhappy lives, which may be necessary for the intensity of their creations or may be a side effect of the extreme intellectual focus.

Beethoven liked to be alone as a child, which may reflect depression. But it adds him to a long list of geniuses, who, even as children, learned to look for their chief satisfactions within themselves and not in the social milieu.

Creativity does not happen in a vacuum. Mozart required a deadline or motivating purpose to compose. How important is the motivating factor, and can there be great creativity without an obvious motivating factor apparent? When commissions were not available, his compositions fell off sharply. This may be true for all genius creative adventures, but it needs to be compared with others in their lifestyles. Genius certainly is accompanied with a keen sense of curiosity, intense focus, and passion! The relative shortness of life for most highly creative persons may have much to do to the temper of their existences with some leading remarkably active lives coupled with both high creative achievements along with high performance demands most of which are self imposed, with Mozart as an example. Others lead relative low-energy or quiet lives, such as Haydn, who had a long life. Both Einstein and Freud had a child with severe emotional problems but were very intellectual, suggesting the alignment necessary for genius in the fathers may be the same alignment in the children with a very different result. Although, a minor but potentially disastrous environmental modification may have created the great difference.

It is quite surprising how much these individuals have in common when their lives are examined with special attention to intellectual energy and intense curiosity, emotional drive, remoteness from peers, and persistence despite apparent absence of motivators other than the intense passion for their individual area of focus. It is equally interesting how different the directions seem to be, despite the apparent common intellectual energy that is likely a function of crucial early modifiers important as to the direction chosen. This obviously raises the key issue of what the outcome is for those individuals with the same basic accelerated intellectual energy, who, for various reasons, never focus or direct their tensions. This will be speculated about later in this work, and the discussion that naturally follows such a conversation includes the potential importance of recognizing the phenomenon early and possibly manipulating key modifiers.

B: CLINICAL SCIENCE

1. INTRODUCTION

Creativity in ECA, as previously mentioned, has defining characteristics that include the ability to produce novel and appropriate work. Individuals with extraordinary cognitive potential armed with this unique neurobiological template at birth, as discussed in the basic science section, then must interface with their unique environments beginning at birth. The environment has a very important influence on their potential in determining the nature and direction of the individual accelerated cognition. It may also be the main determinant for multiple associated behavioral features prominent in their individual lives. Clinical studies of this phenomenon examine the interaction of the underlying cognitive ability interfacing with the environment, searching for patterns of commonality or dissimilarity in group studies to better understand the reasons for the observed remarkable variable outcomes. These clinical discussions will focus on a number of clinical observations made on individuals with ECA, searching for common themes of exposure and behavior and identifying unique outcomes with unique societal exposures.

These areas of clinical study examine multiple aspects of the condition, including the following:

- Familial and peer exposure
- Both early and late environmental influences
- The importance of extreme curiosity from early childhood with omnivorous reading
- Self-education with advancing years
- Early and more permanent psychological issues with neurobehavioral observations Important insights gained from combined studies of creativity and art
- Traumatic brain injuries and anatomical clues to sights of higher cognition
- The role of inspiration and subsequent focus and drive of the creativity
- Clinical measurements providing additional data as to neurobiological origins of ECA
- Review of the various types of thinking that provide the framework or basic tools for creative work.

Numerous group studies will be cited detailing the multiple unique features involved from specific thinking patterns to common psychiatric and emotional struggles prominent in persons with ECA, coupled with general health concerns.

When these clinical studies combine with basic neurobiological information, it becomes clear that the unique contribution of the individual neurobiological template and the unique environmental exposures have very important roles in directing the outcome. Thus, it is no longer an argument of "nature versus nurture," but it is the critical combinations of both influences so important in better understanding the origins of ECA and creativity.

2. CLINICAL SCIENCE DATA

Individuals with the unique neurobiological template thought essential for extraordinary cognitive ability (ECA) potential at birth must then interface with their respective unique environments. This interaction with an infinite variety of potential environmental exposures highly direct the overall outcome for these individuals. Individual and group studies from multiple perspectives provide a new set of important insights for understanding this unique human asset. There may well be an important commonality in the neurobiological template design at birth, but upon engagement with the environment, multiple complex clinical manifestations of this accelerated facility becomes even more informative as to gaining a better understanding of ECA origins. The time-honored debate of nature versus nurture and their respective influences on the origin of extraordinary creativity continues, but most would now agree that both are essential components for the process and play very significant roles in the outcome, as this clinical discussion will further illustrate.

Andrew Steptoe, in *Genius and the Mind,* referred to Mozart's genius and the prodigious performance skills of his sister, Nannerl, makes this point by suggesting an inextricable confound between genetic makeup and early environmental stimulation, allowing champions of both causes to use Mozart as an example (Steptoe 1998).

The following clinical discussion will review the multiple areas of clinical study to better understand the significance of nurture that interfaces with this unique biological template. Immediate environmental influences in general will initiate the discussion, followed by the more specific and focused discussion of key familial influential factors important in developing the creative process. Clinical commentary about the apparent key importance of omnivorous reading for some, the heightened curiosity very early in life, and the role of inspiration and its initiation of creativity, will follow. Multiple, specific types of unique thinking thought crucial for creativity will then be outlined. The relationship of art and ECA provides some interesting insights into the creative process, as do various clinical studies examining individuals with head injury as it affects creativity.

The very important area of affect and psychological profiles and ECA will then be reviewed, followed by a discussion of unique neurobehavioral traits that may be predictive. Multiple clinical monitoring studies will complete the clinical section, with a discussion giving additional information as to the creative process in the clinical setting and utilizing many of the new clinical laboratory techniques informative as to the creative process in vivo.

THE ENVIRONMENT AND CREATIVITY

The environment is a major, if not the major, influential factor in determining the direction and outcome in persons with ECA. Each individual with various levels of ECA at birth must interact with their respective environments, and the environment is so influential during the formative years especially. When one considers how unique we all are at the time of birth in terms of a basic neurobiological template for cognition, coupled with the infinite varieties of potential environmental exposures possible, there is little wonder as to the relative absence of repetitive models with any degree of predictability as it relates to ECA outcome.

Stable, but more important culturally rich homes, loss of parent or parents, financial reversal, and fortune changes all seem to significantly affect the development and outcome of the creative process. Most creators self–educate, and thus scholastic performance does not closely correlate with attainment levels. Juvenile delinquents and depressive or suicidal psychotic patients may exhibit orphan hood rates similar to those of the creative and eminent, so it is clear that such adversity can go either way (Simonton 1999). Thus, the environs can determine the direction and outcome in individuals with ECA, either in a positive individual or societal sense or at times partially responsible for very negative outcomes. Perhaps most importantly, it contributes to the lack of any direction in the majority of these individuals, which may result in multiple varied psychiatric issues instead of productive creative outcomes. Each individual is born with a unique neurobiological template that interacts with equally unique environs, all crucial for individual direction and focus. Familial makeup and exposures, location and timing of location changes, living conditions, schooling, peer influences, early relationships, losses, immediate societal influences, emotional stability or lack thereof in family and peers, and immediate cultural norms during the formative years all have great impact on the individual with ECA.

"Creativity is nurtured," according to Robert Albert (Steptoe 1998). He claims that giftedness, if left on its own, remains at best a potential until it acquires direction and definition. He states that the gifted require specific stimulation and encouragement, and there is little evidence of a genetic influence on creativity independent of intelligence. There may be some debate as to his last point, however. Thus, environmental exposure clearly has a major influential impact on the individual born with the basic neurobiological template responsible for ECA, illustrating the importance of this critical combination of nature and nurture as key promoters for ECA and subsequent creativity with eminence.

THE ROLE OF FAMILY AND ECA

The first question in exploring the clinical data on ECA is to consider what role, if any, the family genealogic background and the immediate family exposure during the formative years has on the outcome of this heightened intellectual ability. Francis Galton, in the mid-nineteenth century, claimed to show evidence of inherited genius from father to son, thus the idea that geniuses are born and not made. The then-British practice of primogeniture and that he included simply well-respected persons as geniuses may have influenced his conclusions. There is now however, no evidence whatsoever for direct inheritance of ECA. Rothenberg and Wyshak, (2004) addressed the question of family background and "genius" by studying fifty Nobel Prize laureates in literature, thirty-one Booker Prize awardees, 135 Pulitzer Prize winners, and twenty National Book and National Book Critics Circle awardees. They also studied 392 eminent persons in noncreative occupations and 560 high-IQ non-prize-winners, plus the general population. They found that outstanding literary prizewinners do not manifest direct inheritance of creativity from their parents; instead, parents and children of the same sex are predominantly in equivalent or performance occupations and have unfulfilled creative wishes.

Eminent persons in noncreative pursuits show up to a 16 percent incidence of same occupation as one or both parents, while that same statistic for creative prizewinners is only 1 percent. This suggests that early developmental influences on child motivation involve identification and competition with the parent of the same sex. Their data would suggest that creative geniuses result from a combination of genetics and environment, both of which may be crucial in the right mix for the process to come to full fruition.

Rothenberg (2005) expanded on the above data by studying the family background of 435 of the 488 Nobel Laureates in chemistry, physics, medicine, and physiology from 1901 to 2003. He compared this group with 548 eminent nonscientists, and there were clearly more correlations in the nonscientific eminent group between individuals and their parents that there were in the highly creative group.

He goes on to restate that there is clearly an early developmental influence on motivation involving identification and competition with the congruent sex parent, which may be a core factor in developing the extraordinary achievement in the chosen field. There are only rare if any examples of direct inheritance of similar focused creativity between parent and child, although highly spirited creativity certainly may be a family trait, often with a different focus. The developmental influence, motivation, and competition in a family certainly may account for the intensity of the focus and specific choice of the field of interest in individuals with accelerated cognition from birth.

It is clear that family influence, especially early in the formative years prior to age twelve, has great impact on the overall direction and intensity of the cognitive energy, if not crucial as to definition of the focus itself and the underlying intellectual drive intensity. It is well understood that these individuals have a unique, heightened curiosity clearly noticeable shortly after birth, which then leads to dependency primarily on self-education for learning. Parental and sibling influences thus have critical roles in allowing this to proceed forward freely or attempting to modify it in different directions. Given the heightened inner tension, unique curiosity, and drive for self-education, the various familial influences are of primary importance as is the accompanying emotional stability of the immediate family.

OMNIVOROUS READING, HEIGHTENED CURIOSITY, SELF–EDUCATION, AND ECA

Extraordinary, unique curiosity very early in life is one very clear repetitive feature in most persons with ECA. This intense curiosity is not only observable very early on but also sustains through the formative years into adulthood. This is coupled usually with a drive for self–education, given an acceptable environment to this approach. Many with ECA are self-driven and sustained in their intellectual curiosity and thus, as a rule, do not have laudable records in school In fact, some have demonstrated difficulty with adaption to school regimens for various reasons often displaying general resistance to school performance.

INSPIRATION AND CREATIVITY

The correlation between inspiration and creativity in persons with ECA is difficult to document in any way other than anecdotal examples. As mentioned previously under environmental influences, several retrospective studies have been done looking at relative large populations of very eminent creative persons, attempting to discover common exposures that seem to be highly correlated with later achievement. Some of this data implies early familial or peer inspiration as contributing factor, but the mix of timing with various relationship exposures makes it very difficult to draw direct correlations. There certainly are some clear examples of the importance of inspiration, but these seem to be related to some fortuitous relationship interactions at a time when the individual is intellectually and emotionally receptive.

Most early inspirational influences seem more important to the personality and underlying drive of the individual, not the specific focus or sector of interest. Clinical studies of parents and role models in the background of highly creative people suggest a possible transmission of high-spirited energy of the parent, but the individual often or most likely predictably chooses another focus for their cerebral energy. It also has been observed that stable, culturally rich homes and loss of parent or parents all seem to have high correlation, along with financial reversal of fortune changes.

SPECIFIC THINKING METHODS IMPORTANT FOR ECA

Multiple, specific types of thinking are deemed important for creative reasoning. These thinking methods allow persons with ECA to be creative with their advanced cognitive processing and may be a key tool important for accomplishing creativity. The content and overall volume capacity of memory, along with the relative ease of accessibility, are key ingredients with ECA. The extent and capacity for visual memory is especially critical. This is called eidetic thinking or the faculty of having perfect visual memory that is easily accessible (Rothenberg 1990). The capacity for "divergent thinking" is very common in these individuals and allows them to pursue two opposed lines of reasoning on the same theme at the same time, a critical thinking method important for new creative conclusions. Multiple opposites or antitheses are conceived simultaneously, and all are equally operative as the thought process moves forward. These cerebration processes help increase the understanding of where true creativity comes from a basic neurobiological standpoint. The Janusian process of thinking describes the ability to equally comprehend multiple opposites or antithesis simultaneously, a key ingredient for creative thinking. Homospatial process thinking is the ability to think simultaneously of opposites in the same space (Rothenberg 1990). The creative mind does not assume "tunnel vision" but is open, questioning, and ready to pursue new ideas in new ways.

"Upside-down thinking" is the opposite of "tunnel vision" and describes a methodology of thinking by looking at an issue or problem initially from a direction 180 degrees from the usual more common approach. "Translogical reasoning," the simultaneous use of varied logical approaches, allows the individual with ECA to approach the problem from a completely new starting point often critical for innovative thought.

Art Insights into ECA

All human activity is ultimately a product of the organization of our brains and is subject to its laws. The study of art reveals the underlying physiology of the brain as an artist becomes a neurologist showing the workings of the brain using a unique technique of study. Creative thinking—especially with the more unique methods discussed in the preceding section—may well be the origin of artistic expression in all its forms. The artist with ECA utilizes these unusual thinking techniques, which leads to creativity in all forms, including artistic expression. Creative individuals have reported auditory, literary, and visual art experiences synonymous with the neurological condition known as synaesthesia, concomitant sensations or a sense other than the one being stimulated. The artistic creations of Alexander Scriabin and Vladimir Nabokov are examples.

Recent evidence shows coherent patterns of neural basis of synaesthesia confirmed with high-spatial resolution brain imaging techniques, and the link with the arts is transpiring to be more than superficial or coincidental. With this apparent clinical "physiology" data these artists and many others provide, an important window into a neural basis of creative cognition. (Mulvenna 2007). A fine line is between creative and destructive thinking, as the personality and art of literary figure Sylvia Plath exemplifies.

AFFECT, PSYCHOLOGY ISSUES, AND ECA

A clear sense of heightened internal tension very early on is apparent in persons with ECA, and it seems to be part of the markedly accelerated cognitive processing. The individual needs to manage this heightened internal tension, and some obviously do better with it than others. This heightened internal tension may provide the exaggerated cognitive energy, which seems so important for the creative process. Unfortunately, some do not manage this internal tension well and various accompanying psychiatric considerations accompany the creative individual. Clinical psychiatric conditions may be apparent, and certain personality disorders are very frequent. Are these disorders necessary for the highly creative process, or are they necessary companions given the primary issue of exaggerated cognitive energy? Clinical schizophrenia has been identified to be often associated with exaggerated creativity (Rihmer, Gonda, and Rihmer 2006). Perhaps with intense focus, a person with ECA becomes highly creative in a positive sense rather than drifting into a clinical schizophrenic state. It is well known that highly creative individuals with ECA historically have a variety of affective disorders, which some have thought were the primary drivers of their achievements and cognitive energy. However, most would consider these affect changes as a predictable personality outcome, given the intensity of constant, highly energized cognitive processing. Affective disorders may therefore be a predictable companion of exceptional creativity.

Creative eminent individuals occupy an unstable terrain between temperament and affective disease (Akiskal 2007). An exaggerated sensibility is found with ECA and to a much lesser degree in persons of talent without ECA. This exaggerated sensibility causes a great part of their real or imaginary misfortunes, with depression and paranoia being common, both of which occur with heightened sensitivity (Lombroso 1891). Hershman and Lieb (1998) have discussed the frequency and intensity of manic-depressive illness in persons of ECA. Some have claimed that manic-depressive illness is a necessary corollary seen in all individuals with ECA. Cyclothymia, without clear clinical manic-depressive illness, seems to be common with heightened creativity (Janka 2004).

In his book, *Origins of Genius* (1999), Dean Keith Simonton lists eminent creative people with identifiable psychopathology. Both Cesar Lombroso (Italy) and Heiophilus Hyslop (Britain) (Steptoe 1998) discussed the fine line between madness and genius and thought that the two often converged in persons with ECA.

Emotion, especially depression, and creativity with various but predictable behavioral aberrations is quite predictable in this group, especially when reviewing known figures with ECA.

CREATIVITY AND BRAIN INJURIES

Clinical head injuries have been shown to influence cognitive processing and may provide some additional insight into the origins of creative ECA. De novo artistic behavior has been indentified following brain injury rarely in persons with frontal temporal dementia, seizures, subarachnoid hemorrhage, and Parkinson's disease. This may provide some insight into the neural basis of creative thinking and human behavior. (Pollak, Mulvenna, and Lythgoe 2007) The clear importance in hemispheric localization may account for this phenomenon, as injuries in the temporal cortex or frontal cortex seem to be the most highly correlated with the posttraumatic appearance of these artistic traits.

NEUROBEHAVIORAL TRAITS AND ECA

Certain common neurobehavioral traits associated with ECA and heightened creativity may be "hardwired" biological differences associated with the basic biological changes responsible for the markedly accelerated cognitive functioning. This suggests that hyperesthesia, unusual or pathological sensitivity of the skin or of a particular sense, (Lombroso 1891) and neuroesthetics, unusual or aberrant brain sensations (Zeki 2001), so commonly associated with ECA may be occurring from originating unique core neuroanatomical and neurophysiologic changes occurring during brain formation that accompany the biological differences and account for the markedly heightened cerebral processing.

NEUROANATOMICAL AND NEUROPHYSIOLOGIC CLINICAL STUDIES AND ECA

Modern clinical measurement technology now provides a clearer understanding of ECA, with the ability to compare measurements of persons with ECA with average individuals. In the clinical setting, this allows recording with clear, reproducible measurements that add to the understanding of the unique nature of heightened creativity. Clinical cerebral blood flow studies comparing individuals with ECA with average controls while performing verbal tasks for the Torrance Tests of Creative Thinking have demonstrated that enhanced cerebral blood flow in certain brain regions, such as right precentral gyrus, right culmen, left and right middle frontal gyrus, right frontal rectal gyrus, left frontal orbital gyrus, and left inferior gyrus and cerebellum, confirm certain selective brain regions critical to the heightened cognitive processing in persons with ECA. This confirms the biological neural basis for creativity and suggests certain neural circuits are hardwired to form the biological structure necessary for ECA.

CLINICAL MEASURES OF ECA

In recent years, there has been a marked expansion of clinical measuring tools of basic intelligence. The Stanford-Binet score was developed at Stanford University in 1918 to improve on the Alfred Binet score, which a

French psychologist developed in 1911, and was primarily designed to test basic intelligence in children.

Now, numerous measuring tools of many types judge the level of intelligence for all ages, primarily with the intent of establishing an intelligence quotient and thus perhaps a prediction of future intelligence performance. These have had some practical importance in assessing the level of intellectual performance and certainly do clearly provide a consistent level of superior performance in individuals with extraordinary cognitive ability. However, they are not that predictive necessarily of creative potential, with the exception of identifying the combination of high intellectual attainment with the ability to utilize the unique thinking methodology described above and thought to be associated with creative ability.

3. Discussion Of Clinical Science Data

The lay term "genius," from the Latin root *Gigno,* "I bring forth," describes individuals who achieve eminence by leaving an impressive body of original and useful intellectual contributions to posterity. Carl R. Rogers defines the phenomena as "emergence in action of a novel relational product growing out of the uniqueness of the individual on the one hand, and the materials, events, people, or circumstances of his life on the other" (Mercado 2008). Genius, assuming both an original and useful outcome from their accelerated intellectual activity, represents the more visible and well known subset of individuals but does not include those with a similar basic neurobiological template for this accelerated facility who do not achieve eminence with either useful or original work. The common feature remains the core neurobiological template responsible for the opportunity of accelerated cognitive facility, and it is in place prior to birth but modifiable over time through the process known as neuroplasticity.

Neuroplasticity is the quality or state of the human brain to be molded, altered, or modified over time by a variety of intellectual challenges operative from day one. These modifications may have significant influence on overall brain function and may play a central part in the early development of the creative potential. Thus, the environment via neuroplasticity has a significant impact on early human brain development and attainment, given all the known key modifiers, including multiple varied early childhood exposures, family relationships, various environmental exposures responsible for development of focused interests, peer influences, location, personality development and intensity of curiosity, mental and physical health, and opportunities.

Genius defines one very important outcome potential from this underlying intellectual process but does not describe all the possible outcomes that occur with accelerated cognitive ability. The immediate family constellation, living conditions, schooling, peer influence, significant family losses, emotional stability or lack thereof, and cultural norms surely are crucial to the critical early environmental influences on brain development. Familial genealogical influence is important to consider, but there clearly is not a one-to-one defining influence for creativity other than perhaps some inherited innate basic motivational energy or cerebral drive that may be a family characteristic.

Clearly, a heightened curiosity associated with early and omnivorous reading coupled with chosen self-education later is a very common pattern in those individuals with ECA. Timely significant inspirational exposures are crucial to the development of the creative process, but the later chosen area for the creativity is usually quite different from the area of interest of the inciting inspiration. The ability to utilize the specific thinking methods outlined in the data section combined with great memory capacity seems to be the key common ingredients with high creative potential. Affect and psychological issues with clinically recognizable personality disorders, such as schizophrenia and paranoia, are common companions in persons with ECA as outlined.

Insight into the creative process has been also obtained by studying de novo artistic performances in persons with frontal or temporal cortical insults. Certain common neurobehavioral traits, such as synaesthesia and hyperesthesia, occur in individuals who are very creative and perhaps signal unusual cerebral processing in combination with the ECA process.

Anatomical and physiological clinical studies have suggested specific cerebral regions or processing that seem to be correlated with the creative process in vivo.

Clearly, a variety of clinically measurable features seems predictive for the highly creative process to flourish and gives us more insight into ECA origins.

4. SUMMARY

Studying common clinical features in persons with ECA or accelerated creative ability provides more insight as to the creative process origin. Numerous environmental elements that prove quite directive in determining the individual outcome of the process influence the unique neurobiological template responsible for ECA potential. Through the mechanism of neuroplasticity, the brain may be significantly modifiable over time and thus becomes a very important driver or determinant of individual intellectual attainment. ECA or genius certainly originates from a delicate balance of genome design coupled with powerful environmental modifiers that largely determine the direction of the cerebral energy, its intensity, and the overall attainment as a result.

PART V:
KEY EARLY MODIFIERS, COMMON CLINICAL FEATURES, AND POTENTIAL IMPORTANCE OF EARLY RECOGNITION OF THE ECA FACILITY

A: IMPORTANT KEY
MODIFIERS

MOST DISCUSSIONS ABOUT ECA ARE APPROPRIATELY directed to better understanding the phenomena as it relates to positive or constructive creative outcomes for society, with multiple time-honored examples throughout the ages. More recently, there has been some interest in this unique phenomenon, which may have been responsible for destructive outcomes or major negative influences on society. The neurobiological template, as previously discussed, responsible for ECA may well be the same underlying biological process but may be responsible for a wide variety of outcomes, ranging from destructive to constructive influences or legacies with a wide range in between, as will be discussed in part 6. The determining factor may well be the presence or absence of certain key environmental modifiers, the timing of the exposure, the receptive status of the individual at the time, and the respective power to influence the focus of the ECA and its intensity. It may well account for the equally important absence of any focus discussed in part 6. It is clear that the individual needs to deal with the constant heightened internal tension from the presence of accelerated cerebral energy, and the path chosen is very sensitive to the key environmental modifiers the individual is exposed to, especially very early in life.

Important key modifiers that play a central role in determining direction and intensity of intellectual focus include the following:

- Early family makeup in combination with a wide variety of cultural and emotional influences
- Persistence and relationships with parents and siblings
- Early loss of a parent, parents, or siblings
- Extended family relationships and the nature of this interface
- Educational exposure, along with the nature of peer relationships;
- Presence or absence of significant mentors or models;
- Perhaps most important, managing the predictable emotional and behavioral issues early that are part of the ECA phenomenon.
- The timing, the intensity and the nature of a wide variety of environmental influences

An obvious potentially great number of complex influences or exposures can be responsible for the unique outcomes that occur in terms of the resultant primary focus, one primary example being how the unique inquisitiveness is managed at a very early age when it is first recognized.

This heightened cerebral energy may be directed in negative directions, the numerous examples of tyrants who had negative impacts on society show. They may have used their intellectual advantage to promote selfish or misdirected agendas. Some have postulated that this specific outcome can be correlated with a manic-depressive affect, which may, in combination with ECA, account for these destructive directions. Are there certain early timely modifiers that promote or direct these individuals in a negative direction? There is no question that these individuals may be significantly intellectually advantaged, perhaps because of this unique neurobiological template that accounts for ECA. But they have become misdirected as a result of certain key, timely environmental modifiers in their early lives. These modifiers are important in considering the intensity and direction of the individual focus and may be central to the absence of any focus, which may result in significant psychiatric disease with its individual and societal burdens.

B: COMMON CLINICAL FEATURES IN PERSONS WITH ECA

CLOSER STUDY OF MULTIPLE INDIVIDUALS WITH ECA who have been highly creative and quite successful with their lives reveals multiple features of commonality that suggest a fair degree of similarity in a number of personal features even though time, space, and environs separate them significantly. This degree of similarity is not surprising and helps confirm a basic underlying commonality of origin in this group, no matter the time or cultural circumstances involved. These features can be identified very early in life and may, when better understood, allow family and peers to recognize the possible presence of the underlying ECA phenomenon earlier.

As initially mentioned in the biography sections, a number of these features reappear in these individuals with varying degrees of intensity. This list is long, but the issues seem constant no matter what the chosen genre is. Some are primary features and others are added characteristics but still contributory. One of the most characteristic features in persons with

ECA is a unique curiosity almost from day one that is intense, continuous, and unrelenting far beyond the norm. Precociousness with an unusually high degree of inquisitiveness seems very common and is associated with a unique, vivid awareness of their environs at a very early age. Most can point to family member or peer association who served as a model or encourager that was important to their development. This is usually associated with a rather obvious heightened "mental" energy and often displays a power to focus beyond the average. Most have a supportive family and extended families with some emotional stability and intellectual stimulation, but this may not be the case in creative individuals who direct their ECA in negatively directed societal pursuits. Loss of a parent or parents is common and may be central in determining the focus and its intensity. Various mood disorders are very common; in fact, some would suggest that to be extraordinarily cognitively advantaged is directly related to some degree of manic-depressive illness. Certainly, depression is quite common, and many of these individuals struggle with emotional insecurity, chronic anxiety disorders, and are very sensitive. Reclusiveness along with suspiciousness is quite common as is a very limited degree of patience for the common and ordinary. Most are not adept socially or even care to be and in general are uncomfortable interacting with others. All seem to have a keen wit, and most are voracious readers from first exposure to written material. Most are usually bored in regular school and do not, in general, do very well academically, especially in the early years. Many, unfortunately, struggle with chronic illness most of their lives, with early death also a common feature. Perhaps this can be related to the extreme internal tension that seems to be a central part of this phenomenon because of the basic neurobiological template, with secondary high, constant cerebral processing. They all have an ability to intensely focus with remarkable perseverance despite medical and personality issues. Usually, no clear data suggests direct inheritable features in families where this phenomenon appears. All these individuals work very hard and therefore expend many hours of intellectual labor, reflecting the degree of internal tension they all feel with the perseverance of relieving, at least in part, some of the inner turmoil. There always is a remarkable disconnect between the power and insightfulness of the inner intellectual processing in comparison with the individual's ability to interact or socially interface with their peers and milieu in general. They are not usually driven by any obvious need for power or material gain of any kind. There may be an additional factor of an inherent intellectual need to satisfy their curiosity beyond just the relief of the exaggerated inner tension and accomplish intellectual goals that set for themselves. They usually express a strong sense of independence and self-confidence with a general intolerance for others, usually are critical of accepted standards in the society, and often become quite progressive in their thinking if not frank reformers.

These common features are present at various degrees, but as a group, they appear constantly with those with ECA. Personality features mentioned along with early and persistent unique curiosity seem universal and I believe a direct reflection of the similar biological template they all are equipped with at birth and must find an acceptable means of dealing with the intense internal pressure that the heightened cerebral energy creates. Having pointed this out, one still must not forget that all individuals are unique. This certainly applies to the ECA subset, but there are interesting similar traits that are perhaps recognizable at an early age.

C: POTENTIAL IMPORTANCE
OF EARLY RECOGNITION

GAINING A CLEARER UNDERSTANDING OF ECA origins and the various clinical manifestations is more than just a passing academic interest of a remarkable and unpredictable human trait. It may well provide very important early signals for recognition and potential insertion of timely modifiers that may significantly influence the outcome with the interactions between the gifted individual and their respective cultures and environments. The core originating determinant may well be a unique, unpredictable alignment of genetic data on the genome, discussed in the basic science section, but when the individual interfaces with the environment from day one, important modifications take place usually that are random. These random modifications have very important implications for gifted individuals and the society in which they are interacting. At the present time, individuals with ECA are not identified early or even later with various peer and education institution interactions. The families, in retrospect, may have been aware of some unique differences versus peers, but it is rare for the basic core attribute to be recognized for what it is and for families and interfacing societies to adjust it. If one accepts that this interfacing is random at present, would appropriate early modifiers lead to more healthy individuals and greater frequency of positive outcomes for society in general? With early recognition, are there opportunities not only to ensure a greater frequency of positive outcomes? More important, can they avoid negative directions of focus? I think there may be a significant number of individuals without any focus at all but who labor through life with tremendous inner tension the unique biological trait generates. They have no form of mental release, and as a result, they suffer from significant clinical psychiatric disease throughout their lives.

If it is true that a significant number of intellectually advantaged individuals in every generation are unable to manage the internal tensions, they may develop significant psychiatric disease secondary to the biological alignment that results in a nervous system tuned to ECA. Early recognition, with appropriate modification of important modifiers surrounding these individuals, might be very preventative and thus positive for both the individual and the surrounding milieu.

PART VI:
ECA WITH A DESTRUCTIVE FOCUS OR WITHOUT A CLEAR FOCUS

A: ECA WITH A DESTRUCTIVE FOCUS

HISTORY IS REPLETE, UNFORTUNATELY, WITH INDIVIDUALS who interact with their peers and environment in a fashion that is destructive to the general good and leaves negative legacies for future generations. Obviously, the origins of this behavior are very complex and may be significantly different from one individual to another. Some of these individuals, thankfully only representing a small minority, display some of the same clinical features as those with clear advantages cognitively and may use their intellectual assets to promote and expand their negative impacts on their contemporary and future societies. These persons are not considered geniuses nor are they necessarily equipped with ECA, given the definitions discussed in part 1, but their accelerated cognitive ability may play a significant role in assisting them in obtaining their perceived goals, in this case with negative individual and societal consequences. These intellectual assets may have the same neurobiological origins as those that meet the clinical definition of genius and leave positive legacies. The extraordinary cerebral activity, whatever that is at birth, may well be very similar to a genius or people with ECA, but because of important environmental exposures with key modifiers discussed earlier, they direct their focus negatively rather or do not develop a focus at all, which can be significant problem in itself. Are there early indicators in their early environmental history that the underlying cognitive gift is not recognized, but other clear modifiers in combination with the intellectual gifts contribute to the unfortunate direction of the focus of the exaggerated cerebral energy?

Much has been written about this in terms of specific clinical origins that lead in these negative directions. Critical early changes in the family constellation, loss of a parent or parents, abuse, rejection, unfortunate milieu exposure, and alignment with individuals or groups outside the cultural standard at the time all may contribute. Would early recognition of the combination of accelerated cognitive ability in combination with these potential negative modifiers lead to potential helpful redirection of focus and result in a much more positive outcome for everyone? Perhaps the most powerful driver and its unfortunate success is when the individual recognizes early that he or she has a special indwelling ability to persuade and lead others with communication skills far beyond the even the most skilled, and these skills can be used to gain personal power.

The accepted definition for genius, as previously mentioned, remains creative work in any genre of endeavor that leads to or results in something new and novel. The basic origin of ECA is most likely neurobiological in origin as stressed in the basic science section. But like most human characteristics or assets, these neurobiological factors are not an all-or-none phenomenon. Thus, there are infinite variations in the intensity of the trait, and innate cognitive ability can occur in all gradations, from a minimal but clear intellectual ability to those with cognitive ability far beyond the average or reaching clinical genius.

It is clear that individuals who have the combination of significant psychiatric diseases—for example, manic-depressive illness along with some intellectual ability—may represent a group in which the focus of one's intellectual advantage may become destructive instead of positive based only on the focus the individual chooses early in life. The mania that may be so destructive in this setting may well be fueled by internalized tension, given the intensity of the intellectual energy in these persons. Some would argue that manic-depressive illness is very common if not essential for ECA or genius.

Napoleon, Adolph Hitler, and Joseph Stalin were tyrants who ruled without mercy and were responsible for egregious crimes against humanity. All three had a severe form of manic depression associated with paranoid delusions and delusions of grandeur if not divinity (Hershman and Lieb 1994). Mania, in this setting, in combination with unlimited confidence, may put the brain into high gear, increasing the speed of thinking and speech and flooding the brain with ideas and energy, culminating in very destructive results. Unfortunately, their significant psychiatric disease was combined with basic cognitive assets and proved a deadly combination, allowing them to gain absolute power.

As an example, Hitler was living in a halfway house in Munich, Germany in the early 1930s. He found, to his amazement, that he was attracting a large crowd of devoted listeners by just speaking passionately on a local political issue of the day. He quickly recognized the potential power of his newly found core asset to verbally communicate his political passion in a very persuasive way and figured out how to use this skill to gain his delusional goals. In retrospect, the most striking clinical feature at this early age and throughout his life was a clear psychiatric disorder of bipolar I manic-depressive illness, which may have been partially fueled by some inherent cognitive advantages (Hershman and Lieb 1994).

He certainly recognized his core asset of verbal communication and used it to his political advantage throughout the rest of his life. He found that he could inject so much passion into his speeches that he could bring the audience easily into his own world and convince them of most anything. Hitler was an angry, lonely, cruel, and dishonest young man who was a dreamer and mystic. He was convinced that he was on an appointed divine mission to restore the Aryan race to its position of superiority after World War I. He had a keen sense of timing, and with his newly discovered core asset of passionate communication, was able to bring the German nation to follow his political agenda. One could certainly argue that this was more about his psychiatric illness than about his perceived intellectual advantages, but early awareness of these important modifiers and their potential impacts—particularly in these individuals—needs to be considered as we learn more about human motivation and the important underlying drivers.

Napoleon Bonaparte and Joseph Stalin, also with significant bipolar I manic-depressive illness, became destructive tyrants and shared many features as Hitler. Again, the apparent psychiatric disease played the major role in their behaviors, as did utilizing intellectual advantage to their benefit in achieving their tyrannical destructive ends.

It would seem that the more we understand the major impacts these early environmental modifiers may have, particularly on identified at-risk individuals, the more important it becomes to attempt to modify the environment toward a more positive focus. The combination of some intellectual ability or awareness may fuel the intensity of a misdirected focus early in life, and thus the early awareness of this potential asset may well be crucial if it is to be accurately identified and potentially directed away from destructive intents.

The following Biography of Joseph Stalin is added to illustrate a potential example of advantaged cognition in a destructive direction.

"It is enough that the people know there was an election. The people who cast the votes Decide nothing. The people who count the votes decide everything"
Joseph Stalin

"Education is a weapon whose effects depend on who holds it in his hands and at whom it is aimed."
Joseph Stalin

B: JOSEPH STALIN BIOGRAPHY

JOSEF VISSARIONOVICH DJUGASHVILI (JOSEPH STALIN)

INTRODUCTION

How did Josef Vissarionovich Djugashvili (Joseph Stalin), raised in the small Russian Georgian town of Gori, become a "mass murderer in addition to an accomplished world statesman responsible for the industrialization of the USSR, organizing Stalingrad for defense during the Second World War, outmaneuvering both Churchill and Roosevelt during their three-power international conferences, and eventually helping defeat Adolph Hitler and Nazi Germany?" (Montifiore 2007) The answer to this question may be a uniquely complicated blend of advantaged intellectual capacity combined with a keen curiosity from birth and, most importantly, coupled with a robust exposure during his early years to a street-fighting culture and lawlessness—all occurring at a time when Russia was primed with revolutionary fervor. This combination of exceptional intellectual curiosity, revolutionary fervor, and especially the behavioral tools he acquired as a youth to survive in a gangster cultural mentality, prepared him for his mission in life: to promote a personal ideal into a real utopia, a mission toward which he proceeded to direct all of his attention, no matter the costs.

"Indefatigable in action, he bubbled with ideas and ingenuity. Inspired by a hunger for learning and an instinct to teach, he feverishly studied novels and history, but his love of letters was always overwhelmed by his drive to command and dominate, to vanquish enemies and avenge slights. Patient, calm and modest, he could also be vainglorious, pushy and thin-skinned, with outbursts of viciousness just a short fuse away" (Montefiore 2007).

Would he have been able to achieve the position of power and leadership he did without his exceptional intellectual abilities and curiosity, which were obvious from a very early age? This relatively unique combination demonstrates in a more contemporary individual the potential destructive consequences of the combination of cognitive advantage with unfortunate early environmental modifiers, which evolved into a very negative societal focus and a passion that was quite destructive to both contemporaries and future generations. Stalin's life and career serve as a frightening example of an advantaged intellect taking a very destructive direction, obviously sensitive to a variety of early life environmental modifiers.

GENERAL BIOGRAPHY

Joseph Stalin was born the third son of Vissarion "Beso" Djugashvili, a cobbler, and Ekaterina "Keke" Geladze Djugashvili on December 6, 1878, in Gori, a small Georgian town in Russia. His given name was Josef

Vissarionovich Djugashvili. His childhood was characterized by jealousy, paternal paranoia, alcohol abuse, infidelity, and domestic violence. The family lost its home, which was Stalin's birthplace, and lived in at least nine different homes plus depressing rental rooms over the next ten years, which was hardly a stable upbringing for Stalin.

He became his mother's "special" person but did not get along well with his father, who was an alcoholic. By the time Soso (a nickname that Stalin carried for a number of years) was five years old, "Crazy Beso" was a severe alcoholic, tormented by paranoia and prone to violence. Some have postulated that the young boy received undeserved beatings from his father. "Beso was violent both to Keke and Soso...'Undeserved beatings made the boy as hard and heartless as the father himself,' believed his schoolmate Josef Iremashvili, who published his memoirs. It was through his father 'that he learned to hate people,'" according to his schoolmate (Montefiore 2007). Stalin learned violence at home and grew up pugnacious and truculent.

At a very early age, it became clear that Stalin was quite different from the norm, but exceptional in many regards. His mother thought he was gifted with clear intellectual curiosity. He read constantly and, early in his life, thought of being a poet or an actor. His performance academically was so advanced that he was later accepted into a Catholic seminary school, where he was exposed to a rigorous classical education—including the world's great literature, with some school restrictions based on content. He performed very well in prayers and all the academic studies, including the Russian language. At age sixteen, he began to write poetry and was the school's most outstanding pupil.

Stalin had the ability to not only navigate successfully with his peers but also be a true leader, which in his youth, unfortunately, was amid a culture of street fighting and lawlessness. His magnetic personality attracted amoral psychopathic personalities around him all of his life, which, when combined with his intellectual capacity, allowed him to become a serious force in the revolutionary movements that he made his utopian societal passion from a very early age.

"All memoirs of his childhood agree that Stalin, even aged ten, exerted a singular magnetism…Somehow, the alternate bullying and crack-up of his father, the passionate adoration of his mother and his own natural intelligence and hauteur, created such a strong conviction that he was always right and must be obeyed that his infectious conviction won him many followers" (Montefiore 2007).

His schooling was quite restrictive in that it allowed only certain written materials and was also a very harsh, regimented, didactic educational experience, perhaps partially explaining why so many ruthless radicals participating in the Russian Revolution came from this same school, the Tiflis Seminary in Georgia. "He knew Nekrasov and Pushkin by heart, read Goethe and Shakespeare in translation, and could recite Walt Whitman… Stalin came to regard himself as a 'special man'" (Montefiore 2007). He read Zola, Schiller, Maupassant, Balzac, and Thackeray, along with Plato, Gogol, Chekhov, Tolstoy, and Dostoevsky. Most of these books were forbidden reading at the Tiflis Seminary, which became a hotbed of political radicals at the time—Stalin, or Soso, being the most prominent.

Stalin owed his political success to the unusual combination of learned street brutality and classical education, which he vigorously pursued all of his life. "His library books are all carefully marked with his notes and marginalia. It was the thoughtful and diligent autodidactic fervour, well concealed under the crude manners of a brutal peasant that his opponents such as Trotsky ignored at their own peril" (Montefiore 2007).

Early on, Stalin decided that his passion was to develop a utopian society through revolution, which would involve a combination of blood, death, and extreme conflict to achieve. "Soso energetically lectured and agitated at his circles. 'Why are we poor?' he asked these small gatherings in workers' digs. 'Why are we disenfranchised? How can our life be changed?' His answer was Marxism and the Russian Social-Democratic Workers Party (the SDs)" (Montefiore 2007). The Russian Social-Democratic Workers Party was founded in 1898, when he first became acquainted with Lenin. The party later split into two factions, the Bolsheviks under Lenin and the Mensheviks under Martov, until 1912, when they totally separated, never to reunite.

Stalin initially caught the attention of the secret police at age twenty-two, when, in the first weeks of 1900, the police arrested him. He participated in the revolutionary demonstration at Baku in January 1905, which turned into the infamous "bloody Sunday." At this time, he had commanded armed men, tasted power, and embraced terror and gangster behavior. He was intimately involved as a revolutionary leader and soon drew the attention of Vladimir Lenin, who led the Communist Party, promoting Marxism. By 1907, Stalin was living the life of a bandit, using espionage, extortion, and agitation as tools of revolutionary success. He became the "godfather" of a small but useful fundraising operation that was like a Mafia family. During these years, he was in prison frequently for subversive antistate activities.

Stalin then emerged as a senior Bolshevik in Russia. In 1917, he was found to be unfit for military service but claimed his leadership at the Bolshevik headquarters. After April 1917, Lenin and Stalin began working closely together, and Lenin considered him to be the "true representative of the rank and file," a true leader of the Caucasians. After the 1917 revolution, Lenin became the supreme ruler of Russia and, on April 3, 1922, appointed Stalin the general secretary of the Communist Party.

After Lenin's death in 1924, Stalin consolidated power for himself, retaining the general secretary position as well as becoming the world leader of Marxism. Considerable political intrigue was involved in the succession from Lenin to Stalin; however, he gained absolute power with his innate leadership and political skills, amplified by his intellectual advantages over all competition, thus eliminating any opposition.

By the late 1920s, Stalin had become the unchallenged leader of the Soviet Union. He replaced Lenin's "New Economic Policy" with his own highly centralized command economy, which was responsible for the severe Soviet Famine of 1932–1933. This economic change also coincided with the imprisonment of millions of people in Soviet correctional camps and the deportation of many others to remote areas. His chronic concern for alleged enemies of his regime resulted in the Great Purge of 1936–1939 in Russia, during which hundreds of thousands of perceived "enemies" were executed, including his old rival Leon Trotsky.

Stalin signed a nonaggression pact with Hitler and Nazi Germany in August 1939, but Germany invaded the Soviet Union in June 1941. He led the defense and turned back the invasion with the historic battle of

Moscow and Stalingrad. He met with both Churchill and Roosevelt during the war to formulate a combined strategy to defeat Germany, as well as with Churchill and Truman after the war. His country was the second nuclear power, and he led the Cold War relationship between the Soviet Union and the United States.

His life consisted of fifty years of conspiracy and thirty years of ruthless government rule. He died at age seventy-four in March 1953. He was responsible for the death of between twenty million and twenty-five million people as he ruled the Soviet Union with absolute, ruthless power.

In retrospect, Stalin was a bright, charismatic psychopath himself from an early age, learning the tools of leadership among gangsters on the streets. He remained an avid reader, became a rebellious seminary student and proclaimed atheist, and then a revolutionary. Throughout his life, Stalin won the devotion of immoral, unbounded psychopaths, which proved to be a core asset for him and which he used for advantage and power the rest of his life.

STALIN AND EXTRAORDINARY COGNITIVE ABILITY

Stalin in his youth had many features of extraordinary cognitive ability, including the key one of robust intellectual curiosity very early in life. In addition, he was a voracious reader, remembered all that he read because of his heightened memory skills, and pursued education aggressively. When he was young, he wanted to be either a poet or an actor. Peers and educators alike encouraged him in his intellectual abilities, and he was accepted into a Catholic seminary school primarily because of those accelerated abilities.

He eagerly responded to this early classical education, reading voraciously in a wide range of subject matter throughout his life. He clearly utilized his intellectual skills to gain power and influence, which came in especially handy for him during his struggle for power and control of post revolutionary Russia in the 1920s and beyond. He had a "detached magnetism" about him, which he discovered to be one of his core assets and used to his personal advantage not only in the street gangster culture of his youth but later at the center of power of the Russian state after 1917.

STALIN: FEATURES OF COMMONALITY AND DISSIMILARITY COMPARED WITH OTHER CREATIVE INDIVIDUALS

Stalin had multiple features in common with other individuals with ECA, including a unique curiosity, voracious reading habits, great memory-retention skills, a capacity for hard work with an extreme ability to focus, self-confidence, strong opinions, and an ability to identify his core personal assets early in life.

Features of dissimilarity are of special interest and may help explain why he took the direction he did, given the negative societal force he became. He was exposed to a variety of very negative environmental modifiers early in his life, including a very abusive alcoholic father, infidelity in his family, domestic abuse, and jealousy, along with the unfortunate early exposure to street crime and revolutionary fervor. Given his intellectual advantage, he soon found his core assets of being a leader with a detached magnetic personality who many others would follow because of his ability to persuade. Another obvious dissimilar feature was how well he performed early in

B: JOSEPH STALIN BIOGRAPHY

school, which was very unusual and may be more a reflection of his underlying sociopathic personality.

STALIN: LEGACY

Stalin's legacy is well known for the magnitude of the destructive focus and direction he took. This focus and direction were born on the streets of Gori in Georgia, inflamed by the unfortunate environmental familial modifiers that were so influential on his underlying psyche. All of this was coupled with an advantaged intellect with great powers of focus and hard work, which serves as a frightening model study of how a clearly negative societal direction can result from the unfortunate combination of intellectual advantage with destruction-promoting environmental modifiers.

C: ECA WITHOUT CLEAR FOCUS

Tracking individual outcomes and secondary societal impacts of persons with ECA with clearly positive or negative focuses are relatively easy with subsequent individual legacies. The individuals who never clearly adopt either a positive or a negative direction with their focus may well compose an important and larger group that most would predict needs much more attention and understanding as to the potential magnitude on individual behavior and societal implications. These individuals may well develop severe personality disorders—if not clear, severe, clinical psychiatric disease later in life—because of attempting to deal with the magnitude of the internal tension they feel constantly. There may well be quite significant and important predictable outcomes without clear direction that are not presently recognized or understood. It is quite clear that persons with ECA are not recognized early in their lives, and thus potentially important societal or cultural modifiers are never in place to attempt to provide positive directions for this unique neurobiological attribute. As we better understand ECA origins and the key environmental modifiers as to directing the heightened cerebral energy, we may not only influence focus in a positive direction but also address the equally important and unfortunate outcome of those who had ECA but never proceeded in either a positive or negative direction.

Unique DNA alignment designated emergenesis as previously discussed may be the core initiating factor ECA appearance with perhaps markedly accelerated neuronal conduction velocity and marked expansion of neuronal networks or connections. This basic neurobiological attribute creates massive internal tension that the individuals must confront as they engage society, as previously discussed. All persons with ECA are running "red hot" from a cerebral standpoint, creating massive internal tension that must become overwhelming at times, especially if cognitive pursuits are random without direction. Thus, it is clear that very early environmental modifiers become extremely important in how an individual handles this increased internal tension and begins to develop cognitive focus and direction. These individuals develop unique personality traits very early on, such as impatience and self-centered attitudes. They are highly inquisitive, and family and peers alike perceive them as being quite different from day one.

The key question is attempting to define what is responsible for and what drives the lack of focus in these individuals. A focus pursued intently appears to relieve the internal tension that is so extreme in these individuals. Even with clear focus, significant personality disorders are common as the intensity of the internal tension remains high.

What prompts an individual with ECA to move in a negative or destructive direction? Is there some early parental or other family and peer abuse without awareness of the special attribute of the individual? These individuals are not "normal" and family and peers perceive them as "odd" or different. Thus, they may be ignored, belittled, or even bullied. Families—especially parents may—become abusive and hostile with the inability to understand their differences or control the young person with ECA. They may often become involved in a very hostile environment.

What happens with these individuals when they develop neither a positive nor a negative focus? The answer is not clear but one could speculate that they have serious problems engaging their environments and society, which may result in significant clinical psychiatric disease. This may include significant personality disorders, manic/depressive disorders, frank psychotic disorders with suicide potential, and schizophrenia. They may develop maladaption syndromes with family and peers, which results in additional engagement injury because of the lack of understanding of the origin of their uniqueness. If all this is true, it may then become even more important to recognize the basic neurobiological source driving the behavior, ECA, to allow for possible earlier understanding and potential modification of important environmental markers that may be beneficial to the individual.

A significant number of individuals may be identified with significant psychiatric disorders. These disorders may, at least in part, be attributed to an underlying, profound anxiety disorder secondary to a constant accelerated cerebral processing disorder, ECA, and are not relieved to any degree because of the lack of a passionate focus or intellectual direction. Even persons with clear clinical genius or ones with clear destructive focuses are overwhelmed at times with the degree of internal tension they feel as part of the accelerated cognition from birth. It would not be a big surprise to discover a substantial population with significant clinical psychiatric disease and with magnified inner tension and an overwhelming anxiety disorder that's partially attributed to ECA. If this is true, early recognition and important key modification with the early recognition becomes crucial in potential prevention.

PART VII:
LEGACIES OF ECA

A: INTRODUCTION

THIS PART WILL DETAIL THE SOCIETAL interface and overall legacies of groups or subsets of individuals who are clearly focused with a strong direction intellectually, either in a positive or negative direction, in terms of societal impacts to complement the discussion about individuals without a clear focus discussed in part 6. This discussion may well encourage earlier recognition of this unique neurological trait and the critical role of important factors and their timing.

Clearly, individuals with ECA are equipped with this unique physiological and anatomical biologic template at birth, and it may have extraordinary impacts on their own milieu and future generations. It is well accepted that the work of individuals with this unique cognitive advantage often result in remarkable creativity and invention. This in turn results in a very positive legacy for society and is usually referred to as the phenomena of genius. There are multiple examples throughout the ages. Others with the same clear ECA may have legacies or long-term societal impacts that may not be positive. This discussion may well encourage earlier recognition of this unique neurological trait and the role of important modifiers and timing that may play in the overall outcome in individuals with ECA or an intellectual advantage from birth.

B: POSITIVE FOCUS WITH CLEAR, LASTING IMPACT ON SOCIETY

THROUGHOUT HISTORY, A NUMBER OF INDIVIDUALS identified as persons of genius with remarkable cognitive ability have had positive impacts on their contemporary societies and left profound legacies for future generations. It seems clear that ECA is the core common feature and fortunately is directed in positive directions the peer society and for future generations to come. Historical examples of this have been provided in part IV, with individuals from a variety of pursuits including music, literature, and technology and leadership. There are many other obvious examples throughout history with intellectual pursuits in other directions including political leadership, education, philosophy, science, medicine, law, and business, to identify only a few. These individuals all have ECA and must therefore deal with the accompanying inner tensions that seem to be a central part of the phenomena itself. This may account for the problems that all these individuals seem to have interfacing with society with recognizable personality disorders, if not frank psychiatric disease for some. It is as if the intellectual asset is both a gift and a curse, as suggested in the title of this book.

These persons are "difficult" to relate to and lead very difficult lives with problems interfacing with their respective societies and adjusting to the intense inner tension, if not having major medical issues and shortened lives because of the multiple hazards they encounter. They are not part of the mainstream, and their peers perceive them as different. There are two additional aspects important to point out particularly about individuals with positive legacies with ECA as likely the core attribute. The first is how very difficult it must be for these individuals to navigate through the emotional hazards central to the phenomena, along with the significant problems with peer social interaction that seem to be a constant. Yet they achieve despite these mountainous hurdles. The second is attempting to understand the underlying motivational force that seems so strong in the absence of obvious materialistic gain or personal power, which seems so important for the intellectually ungifted. Is part of the genetic predisposition, along with the unique genome configuration, a powerful central need to perform robustly intellectually and thus the process itself is a response to a predetermined genetically designed need?

Most of these individuals have left important legacies for future generations but largely have paid a severe personal price to achieve what they did. The world has clearly greatly benefited from the remarkable contributions that most of them have made, primarily from personal creativity leading to political, societal, and technological advances in combination with the great gifts of art that we all continue to enjoy today.

C: NEGATIVE SOCIETAL FOCUS WITH IMPLICATIONS

INDIVIDUALS WITH THIS SAME UNIQUE NEUROBIOLOGICAL template with outcomes that are not positive for society and actually lead in a negative or destructive direction with unfortunate impacts on contemporary societies and destructive legacies for future generations is less understood. These individuals are equipped with presumably a similar neurobiological template as persons with ECA who were positively directed at birth, but for a variety of reasons, these individuals chose a different direction for their focus, perhaps responding to a combination of key environmental modifiers that influenced attention in a destructive direction. It is difficult to know what the role of accelerated cognition is for these individuals in general, and it is doubtful that any of these persons have extraordinary cognitive agility or genius by the definitions previously discussed. However, behind the obvious personality disorders if not frank psychiatric disease, especially manic-depressive disease, enhanced cerebral energy in the form of rapid thought, accelerated memory, intense focus for long hours, and high degrees of self-confidence suggest some components of accelerated cognitive ability. It is doubtful that the advantaged cognitive ability is primary in these individuals, as the psychiatric variations seem so obvious. But some degree of cognitive advantage may well play a role when combined with manic-depressive disease, for example. This certainly may well have been the case for Joseph Stalin, Adolf Hitler, and Napoleon. All three had several of the common intellectual features previously discussed associated with ECA, but no one would want to call them geniuses, given their obvious psychiatric issues and destructive behavior. Did they have similar neurobiological templates at birth, but given their personalities and exposure to certain environmental modifiers, they became very destructive with their behaviors? It raises the issue again of the importance of early recognition of this accelerated cognitive condition in individuals with significant psychiatric problems for potential positive modification and hopefully redirection of the pent-up energy in a positive societal direction rather than being allowed to drift into very negative directions.

Some of the key early modifiers that may be critically important to identify in leading to negative pursuits in combination with some degree of ECA include rejection, clear psychiatric problems, abuse, parental loss, major family dysfunction, environmental hardships, financial reversals, relationship problems with peers, and major political/social disruptions and many others. See related material from Hershman (1994).

Would there be an opportunity to constructively intervene if one were able to identify the apparent critical combination of ECA with potential destructive modifiers listed above? The combination is easy to recognize in retrospect but may not be workable if recognized in the early development of the personality.

D: INDIVIDUALS WITHOUT CLEAR FOCUS BUT WITH ECA

PERHAPS OF EVEN MORE CONCERN IS the less well understood or at least infrequently discussed potential societal impact of those individuals with ECA who never focus their heightened cerebral energy in either a positive or negative direction and thus may evolve a completely different set of problems for society, as discussed previously in part 6. These individuals, I believe, are equipped with the same neurobiological framework at birth that has been discussed for individuals with ECA who become creative and may deserve the clinical label of genius. The major variation, however, seems to be the absence of a focus, which I believe is crucial to successfully dealing with the intense internal cerebral energy so apparent in these individuals almost from day one. The basic neurobiological template at birth for ECA clearly is associated with intense internal tension that I think is, at least in part, a function of exaggerated neuronal conduction velocity with heightened cerebral energy in general. Without a focus, the individual must deal with this markedly elevated internal tension and therefore does not find an outlet or safety valve to relieve the pressure. This may lead to significant primary psychiatric issues that become a serious adaption problem for the individuals, their families, and society in general. Previously, the issue of a thin line between madness and genius and the clinical observation that a great percentage of highly creative people have significant depressions and many has apparent manic-depressive syndromes. One is lead to speculate that there may be a significant chance that an individual with ECA who doesn't have a clear outlet for heightened cerebral tension develops clinical psychiatric disease in the form of serious clinical mood disorders if not full-blown psychotic disorders. This would be difficult to prove clinically without a double-blind study, which would be very difficult to do, if not impossible. I believe there is now enough evidence that very early identification of the multiple common clinical features discussed elsewhere in this work and very early appearance in these individuals with ECA who clearly do not have an obvious focus for their enhanced cerebral energy is all the more important. With very early recognition of the ECA trait followed by awareness of the key modifiers, perhaps this potential group of psychiatrically impaired individuals, occurring partially as a result of their heightened cerebral energy, could be positively modified.

PART VIII:
CONCLUSIONS AND
SUMMARY

A: DISCUSSIONS AND CONCLUSIONS

Specific neurobiological template changes, including unique anatomical, chemical, and physiological configurations likely genetically determined by a process of emergenesis as previously discussed provides the core substrate that allows the opportunity of very high cognitive functioning. This substrate pattern developed with critical and unique neurobiological combination complexes does not guarantee creativity or genius, but it may be the necessary structural framework to allow for the possibility given the right environmental exposure in combination with certain key personality traits, such as high energy, motivation, ability to focus, visual imagery retention, mood swings etc.

It is important to point out that Lombroso (1891) emphasized gross anatomic changes in the "celebrated," including skull capacity and cortex organization, which suggests anatomical differences that are not probably correct, at least at the macroscopic level. The differences may well be microscopic, however, in neural network organization, secondary Darwinism, synaptic velocity, and neurotransmitter concentrations. The proverb "a man who has genius at five is mad at fifteen" leads to the whole discussion of what happens with this critical neurobiological substrate is in place but lacks environmental support or nurture. Creativity does not become manifest. What happens to these individuals with all this potential unique ability or talent that becomes misdirected?

These persons with accelerated cognitive ability from birth are driven to release their inner tensions and thus are driven to create. They pay a terrible price for this high-velocity cerebral energy, as they are not able to lead normal lives in the sense of fitting in well with their contemporaries. This unique cerebral advantage, present at birth and colored by environmental influences, works well for their creative needs, as it relieves their inner tensions. However, it leaves them with a mental profile that creates considerable discomfort in a social sense and often is complicated by both physical health problems and psychiatric complications, such as severe mood disorders, bipolar disease, paranoia with heightened sensitivity, and generally shortened lifespans. Some say that all of these persons have some degree of a bipolar disorder. What is responsible for the primary focus of their creativity? Is it in the genes, or does the environmental exposure drive it? All seem to have a precocious ability that is easy to recognize early on. The role of both physical and mental disorders looms large in all their lives. They all have a very characteristic, keen sense with recognizable high-cognitive ability and a great need to understand everything that they meet. A high degree of curiosity, coupled with omnivorous reading, adds constantly to the richness of their experiences. They all seem to be insatiably inquisitive and approach all engaging mental activities with high intensity, and for the most part, are all self-educated.

Biographic analysis of persons with very high cognitive ability seems to suggest that a unique basic neural network is the necessary foundation for accelerated cerebration and is obvious from birth. This unique neural network made up of unusual anatomical, physiological, and chemical features accounts for the supercharged mental energy. This framework most likely develops from a critical combination of DNA determinants, perhaps involving emergenesis in those persons without obvious preceding similar attributes in family members of preceding generations.

In addition, these individuals have a genetically determined trait identified as neuroplasticity, which accounts for the remarkable ability to form new neuronal pathways throughout life, greatly enhancing their already greatly accelerated cognitive abilities. This chance genetic combination or emergenesis may lead to the basic underlying framework necessary for the genius process to become manifest. It is necessary to clearly define what we mean by the term genius if we are to begin to decide as to the relative importance of the multiple possible underlying modifiers that may be important to the phenomenon. There is considerable disagreement as to the most appropriate definition, depending on what viewpoint you are coming from. The neurological behavioral and historical approach focuses on outcome and recognition of the creativity as the ultimate benchmark of genius; however, the basic scientist would argue that the genius phenomena might occur without ultimate recognition or success because it is a basic biological process that occurs and proceeds whether the individual ever is identified as being creative or successful. Measures are difficult and go well beyond the standard IQ test, which may be correlated to creativity. The most important measure may well be one of simply synaptic velocity, neural conduction velocity, memory storage, and accompanying intensity of thought. This may explain the heightened awareness and focus easily recognized very early in the process, which may lead to success or failure in terms of societal acceptable human achievement and always seems to be associated with a clear discomfort level when attempting to navigate the mundane. Understanding the multiple facets of genius should allow us to better understand standard mechanisms in human cognition. The importance of nurture is obvious, as this heightened velocity of thought is met by the specific environment in which it finds itself and thus must interact and proceed as influenced by the specific nature of the environmental exposure. This is especially critical in the very early family milieu exposure and parental interaction. No two environments are the same and by definition are very complex. Thus, they may modify the basic heightened energy in many different ways. This may be a crucial influence as it relates to direction, focus, degree of discomfort for the individual, success, or failure, and thus be either a positive or a negative legacy for future generations. It may well be that severe, destructive directions may occur with the basic genius framework in place when environmental forces allow accelerated cerebral energy to become a perceived negative focus and a resultant negative legacy, perhaps as a function of unawareness along with the loss of childhood malleability after the age of twelve. This idea has not been explored extensively in the literature but should be as to better understand the legacies and the initial direction and focus that account for it. These observations and understandings may have important implications on our education programs going forward. Perhaps more important, it has implications for better understanding what constitutes healthy environmental exposures early in life. Positive legacies are of much greater interest and can be studied in the many varied biographies of individuals throughout history. The impact of these legacies is of great interest to us all as future generations speculate as to the importance of these legacies and what our lives would have been without influence that was a result of this basic brain process.

Destructive legacies also need to be explored because they are not only fascinating but may have serious practical implications for society in general as we might learn to better recognize the basic energy of ECA early and understand how external influences that might have been present to alter the direction and the intensity of the energy. There may be important tools here for both educational and early family and peer relationships to understand how best to proceed or at least to know what to avoid.

B: SUMMARY

"IN EVERY WORK OF GENIUS, WE recognize our own rejected thoughts: they come back to us with a certain alienated majesty" (Emerson 1847). "Self Reliance" was a central theme in Emerson's works and underlies a critical human asset important in understanding accelerated cognitive ability in man. As to the origin of ECA, Ackroyd's Charles Dickens may have described the potential origin the best in terms of the unique emergenesis theory by remarking; "Of course in the son it was controlled and magnified to an unimaginable extent, but the mystery of inheritance remains. And it is a mystery how such gifts and qualities can be transmitted from generation to generation, until there comes a time when in one particular person they blaze out and, in a sense, devour all those around" (Ackroyd 1990). People with this gift have a unique configuration of their neurobiological frameworks, which include neuroanatomical, neurophysiological, and neurochemical designs that represents the core essence of this unique cognitive ability.

This extraordinary basic neurobiological template likely provides markedly accelerated processing with greatly increased neuronal transmission speed, resulting in remarkable memories with attendant keen sense of focus, extreme curiosity, and ability to process multiple thought process simultaneously associated with a basic cerebral energy level far beyond the norm. This is the basic feature of this process and must be in place before birth to allow the later development of the phenomena. This unique neurobiological framework is understood as an unusual alignment of genetic materials that may provide for an altered neurological functioning that the individual must begin to deal with very early in life. There may well be important structural network configurations or hard wiring considerations as well in originating this process beyond just altered brain anatomy and neurophysiology. This phenomenon is manifested very early in life with very high cerebral energy, unusual inquisitiveness, difficulty conforming to the usual, and exceedingly high powers of perception and enhanced memory retention. These individuals need to adapt to this high cerebral energy and attempt to understand it without many around them of similar ability. This leads to adaptation problems and persistent maladaptation issues that go on for the life of the individual. These psychological issues are not primary to the phenomena but are expected secondary effects to the underlying uniqueness of the cerebral design. We can understand the personality features and attempt to develop insight into the process through many psychological studies; however, I believe the psychological issues are secondary to the primary process. This would be expected, given the struggle these people have in attempting to adapt this "gift" to their surroundings.

The biographical studies, therefore, are of great interest in terms of studying the process of higher cognitive development. But focusing too much on the behavioral issues will not help us understand the process better, other than displaying the expected and multiple problems that occur with adaptation. The product of this accelerated cognitive functioning and its legacy for generations to come is of great interest as the result of these phenomena and is part of any biographical study. But it may not lend much to the underlying understanding of the phenomena itself. There is the question of the negative results of this unique process in situations where the basic neurobiological framework is present but for whatever reason, the focus or direction of the energy becomes misdirected and results in a negative legacy for generations to come. This process needs to be better understood early as a remarkable subservice neurobiological structure allowing for ECA potential may exist but the resulting surface behavioral aberration issues that may be predictable need to be identified and managed. Identification early in life, along with the key modifiers, allowing positive focus, allowing expected unusual behavior, may be important in the future to better maximize the gift and perhaps prevent tragic, negative directions. Preventing negative directions for those of genius endowment also becomes a great challenge and may be more important in the long run. An equally important issue is for the society of those individuals who possess this basic neurobiological framework but do not focus in either a recognizable positive or negative direction but relieve their inner cerebral tensions in ways that may explain the origins of more common psychiatric or behavioral disorders that society has and must deal with.

This study has outlined the basic core neurobiological framework necessary for this process to occur and outlined numerous individual clinical studies to better understand how these persons operate in the environments that they find themselves. Numerous biographical studies have shown the phenomena on an individual basis over the years, which demonstrate many common themes. These features include social adaptation problems, very early unique curiosity, struggles with the high cerebral energy and how to cope with it, misunderstanding by their peers, impact of major illness in their lives, process of focus and energy expelled to their products, and either their great important and positive legacies or great negative forces for future generations. The biographic point may be that the intense internal drive or focus propels the inner life, and the external life that accompanies it is beside the point, as it can be anything but is predictably chaotic given the intensity of the inner focus.

There is no question that this ECA phenomenon has greatly impacted human history in so many ways, and as we learn more about its origin, including the basic neurobiological template in combination with the key environmental modifiers, we may develop an opportunity and ability to recognize its existence early and allow positive modification of the direction of its power. Possibly, we can prevent serious psychiatric disease in those who do not successfully find an outlet for the exaggerated internal tension that underlies the phenomenon.

PART IX:
REFERENCE LIST

Ackroyd, P. 1990. *Dickens*. New York: Harper Collins Publishers.

Akiskal, J. S., K. K. Aliskal, and S. Haqop. 2007. "In Search of Aristotle: Temperament, Human Nature, Melancholia, Creativity, and Eminence." *Journal of Affective Disorders* 100(1):1–6.

Ashoori, A. and J. Jankovic. 2007. "Mozart's Movements and Behavior: a Case of Tourette's Syndrome?" *Journal of Neurology, Neurosurgery, and Psychiatry* 78:1171–1175.

Baker, C. 1968. *Ernest Hemingway: A Life Story*. New York: Avon Books A Division of the Hearst Corporation

Bedford, H. 1925. *Robert Schumann: His Life and Work*. New York: Harper and Brothers.

Bloom, H. 2002. *Genius*. New York: Warner Books.

Brain, D. 1996. *Einstein: A Life*. New York: John Wiley & Sons.

Braunbehresn, V. 1986. *Mozart in Vienna 1781–1791*. New York: Grove Weidenfeld.

Brooks, V. W. 1932. *The Life of Emerson*. New York: E. P. Dutton and Co.

Bull, G. 1998. *Michelangelo, a Biography*. New York: St Martin's Griffin.

Charles, M. 2004. "The Waves: Tensions between Creativity and Containment in the Life and Writings of Virginia Woolf." *Psychoanalytic Review* 91:71–98.

Charlton, B. G. 2009. "Why are modern scientists so dull? How science selects for perseverance and sociability at the expense of intelligence and creativity." *Medical Hypotheses* 72(3):237–243.

Chavez-Eakle, R. A., A. Graff-Guerrero, J. C. Garcia-Reyna, V. Vaugier, and C. Cruz-Fuentes. 2007. "Cerebral Blood Flow Associated with Creative Performance: A Comparative Study." *NeuroImage* 38(3):519–528.

Chklovskii, D. B., B. W. Mel, and K. Svoboda. 2004. "Cortical Rewiring and Information Storage." *Nature* 431(7010):782–788.

Colombo, J. A., H. D. Resin, J. J. Miguel-Hidalgo, and G. Rajkowsda. September 2006. "Cerebral Cortex Astroglia and the Brain of Genius. A Propos of A. Einstein's Brain." *Research Reviews* 52(2):257–263.

Davenport, M. (1932) *Mozart.* New York: Barnes and Noble Books.

Davis, P. 1998. *Charles Dickens A to Z: The Essential Reference to His Life and Work.* New York: Checkmark Books.

Dietrich, A. 2004. "The Cognitive Neuroscience of Creativity." *Psychonomic Bulletin Review* (6): 1011–1026.

Doerr-Zegers, O. 2003. "Phenomenology of Genius and Psychopathology." Psychiatria et Neurologia Japonica 105(3):277–86.

Durant, W. and A. Durant. 1967. *Rousseau and Revolution.* New York: MJF Books.

Ebner, F., Neubauer, A. C. 2008. "The Creative Brain: Investigation of Brain Activity During Creative Problem Solving by Means of EEG and FMRI." *Human Brain Mapping,* Published online 11 Feb. 2008.

Ehrenwald, J. 1984. *Anatomy of Genius, Split Brains and Global Minds.* New York: Human Sciences Press.

Emerson, R.W. 1847. Self Reliance. Major American Writers Third Edition. New York: Harcourt, Brace and Company.

Ewen, D. 1931. *The Unfinished Symphony, a Story of the Life of Franz Schubert.* New York: Modern Classics Publishers.

Fields, R. D. 2004. *Scientific American,* April: 54–61.

Fink, A., M. Benedek, R. H. Grabner, B. Staudt, and A. C. Neubauer. 2007. Creativity Meets Neuroscience: Experimental tasks for the Neuro Scientific Study of Creative Thinking. Methods. 42(1):68–76.

Fink, A., R. H. Grabner, M. Benedek, G. Reishofer, V. Hauswirth, M. Fally, C. Neuper, F. Ebner, A. C. Neubauer. 2008. "The Creative Brain:

Investigation of Brain Activity During Creative Problem Solving by Means of EEG and FMRI." *Human Brain Mapping* 30(3): 734-748.

Flaherty, A. W. 2005. "Frontotemporal and Dopaminergic Control of Idea Generation and Creative Drive." *The Journal of Comparative Neurology* 493(1):147–153.

Flower, N. 1928. *Franz Schubert, the Man and His Circle*. New York: Frederick A. Stokes Company.

Garcia, E.E. 2004. Rachmaninoff and Scriabin: "Creativity and Suffering in Talent and Genius." *Psychoanalytic Review* 91(3): 423-442

Gardner, R. J. M. 1988. "Emergenesis? If not, What?" Letter to the Editor. *Am J Hum Genet* 43 (3):344.

Gardner, H. 1993. *An Anatomy of Creativity Seen Through the Lives of Freud, Einstein, Picasso, Stravinsky, Eliot, Graham, and Gandhi*. New York: Basic Books, A Division of Harper Collins Publishers.

Gazzaley, A. and M. D'Eposito. 2007. "Part II: In Vivo Imaging of Human Aging and the Transition to Cognitive Impairment. Top-Down Modulation and Normal Aging." *Annals NY Acad. Science* 1097:67–83.

Geiringer, K. Haydn. 1963. *A Creative Life in Music*. Garden City, New York: Anchor Books, Doubleday and Company.

Goldberg, T. E. and D. R. Weinberger. 2004. "Genes and Parsing of Cognitive Processes." *Trends in Cognitive Sciences* 8(7):325–335.

Goldsmith, B. 2005. *Obsessive Genius, the Inner World of Marie Curie.* New York and London. W.W. Norton Company

Goodwin, D. K. 2005. *Team of Rivals, the Political Genius of Abraham Lincoln.* New York: Simon & Schuster.

Goulding, P. G. 1992. *Classical Music: The 50 Greatest Composers, and Their 1,000 Greatest Works.* New York: Ballantine Books.

Heilman, K. M. 2009. Creativity and Aging: Presented as Course Material at the American Academy of Neurology Course on Creativity and Neurologic Disease May 1, 2009. 237–243.

Heilman, K. M., S. E. Nadeau, and D. O. Beversdorf. 2003. "Creative Innovation: Possible Brain Mechanisms." *Neurocase* 9(5):369–379.
Hershman, D. J. and J. Lieb. 1994. *A Brotherhood of Tyrants: Manic Depression and Absolute Power.* New York: Prometheus Books.

____ 1998. *Manic Depression and Creativity.* New York: Prometheus Books.

____ 1998. *The Key to Genius.* New York: Prometheus Books.

Heywood, R. B. (ed.) 1947. *The Works of the Mind.* Chicago, Illinois: The University of Chicago Press.

Hotchner, A. E. 1955. *Papa Hemingway, A Personal Memoir.* New York: Random House.

Irving, W. 1978. *The Legend of Sleepy Hollow and Other Stories (The Sketch Book of Geoffrey Crayton, Gent)* New York: Penguin Books.

Isaacson, W. 2007. *Einstein, His Life and Universe.* New York: Simon & Schuster.

___ 2011. *Steve Jobs.* New York: Simon & Schuster.

Jacob, H. E. 1950. *Joseph Haydn, His Art, Times, and Glory.* London: Vicor Gollancz LTD.

Janka, Z. 2004. "Artistic Creativity and Bipolar Mood Disorder." *Orv Hetil.* 145(33):1709–1718.

Jung, R. E., R. S. Charles, R. A. Chavez, S. M. Flores, S. M. Smith, A. Caprihan and R. A. Yeo. 2009. "Biochemical Support for the Threshold of Creativity: A Magnetic Resonance Spectroscopy Study." *The Journal of Neuroscience* 29(16):5319–5325.

Jung,R. E., J. M. Segall, H. J. Bockholt, R. A. Flores, S. M. Smith, R. S. Chavez, and R. J. Haier. 2010. Neuroanatomy of Creativity. Human Brain Mapping 31(3): 398-409.)

___ 2010. "Neuroanatomy of Creativity." *Human Brain Mapping.* 31(3):398–409.

Kaplan, F. 1988. *Dickens, A Biography.* New York: Avon Books.

Kiehl, K. A. 2006. "A Cognitive Neuroscience Perspective on Psychopathy: Evidence for Paralimbic System Dysfunction." *Psychiatry Research* 142(2-3): 107–128.

Landrum, G. N. 1993. *Profiles of Genius, Thirteen Creative Men who Changed the World.* New York: Prometheus Books.

Lanni, C., S. C. Lenzken, A. Pascale, I. Del Vecchio, I., M. Racchi, F. Pistoia, and S. Govoni. 2008. "Cognition Enhancers between Treating and Doping the Mind." *Pharmacological Research* 57(3) 196–213.

Larned, J. N. 1911. *The Study of Greatness in Men.* New York: Houghton Mifflin Co.

Leonard, R. A. 1946. *The Stream of Music: Chopin.* Garden City, New York: Doubleday.

____ 1946. *The Stream of Music: Beethoven.* Garden City, New York: Doubleday.

____. 1946. *The Stream of Music: Schubert.* New York: Doubleday.

Li, C. C. 1987. "A Genetic Model for Emergenesis: In Memory of Laurence H. Snyder, 1901–1986." *Am J Hum Genet* 41(4):517–23.

Lieb, J. 2008. "Two Manic–Depressives, Two Tyrants, Two World Wars." Medical Hypothesis E Published 70(4): 888-892

Lombroso, C. 1891. *The Man of Genius.* New York: The Walter Scott Publishing Co. LTD. and E. C. Charles Scribner's sons Publishers.

Ludwig, A. M. 1995. *The Price of Greatness, Resolving the Creativity and Madness Controversy.* New York: Guilford Press.

Lykken, D. T. 1982. "Presidential Address, 1981. Research with Twins: The concept of Emergenesis." Psychophysiology 19(4):361–373.

Lykken, D. T. 2006. "The Mechanism of Emergenesis." *Genes Brain Behavior* 5(4):306–310.

Lykken, D. T, M. McGue, A. Tellegen, T. J. Bouchard Jr.1992. "Emergenesis: Genetic Traits that May Not Run in Families." *Am Psychol.* 47(12):1565–1577.

Marek, G. R. 1985. *Schubert.* New York: Viking Penguin.

Mesulam, Marsel M. 2000. *Principles of Behavioral and Cognitive Neurology.* Second Edition. New York: Oxford University Press.

McAllister, A. K. 2007. "Dynamic Aspects of CNS Synapse Formation." *Annu Rev Neurosci* 30:425–450.

McCullough, D. 2001. *John Adams.* New York: Simon & Schuster.

Mercado, C. 2008. "Neural and Cognitive Plasticity: From Maps to Minds." *Psychology Bulletin* 134(1):109-137.

Montefiore, S. S. 2007. *Young Stalin*. Great Briton: Weidenfield & Nicolson.

Moore, D. W., R. A. Bhadelia, R. L. Billings, C. Fulwiler, K. M. Heilman, K. M. J. Rood, and D. A. Gansler. 2009. "Hemispheric Connectivity and Visual-Spatial Divergent-Thinking Component of Creativity." *Brain and Cognition* 70(3):267–272.

Mulvenna, C. M. 2007. "Synaesthesia, the Arts and Creativity: A Neurological Connection." *Front Neurol Neurosci* 22:206–22a.

Murdoch, W. 1935. *Chopin, His Life*. New York: The Macmillan Company.

Murray, P. (ed.) 1989. *Genius, The History of an Idea*. Great Britain: T. J. Press.

Neumayr A. 1994. *Music and Medicine Haydn, Mozart, Beethoven, Schubert: Notes on their Lives, Works, and Medical Histories*. Bloomington, Illinois: Medi-Ed Press.

Noyes, G. R. 1918. *Tolstoy*. New York: Dover Books.

Oerter, R. 2003. "Biological and Psychological Correlates of Exceptional Performance in Development." *Ann. N.Y. Acad. Science* 999:451–460.

Payne, R. 1995. *The Life and Death of Adolf Hitler*. New York: Barnes & Noble Books.

Pennisi, E. 2006. "Neuroscience. Brain Evolution on the Far Side." *Science* 314(5797):244–245.

Pollak, T. A., C. M. Mulvenna, M. F. Lythgoe. 2007. "De Novo Artistic Behavior Following Brain Injury." *Front Neurol Neurosci* 22:75–88. (Full Text Not Available).

Popp, A. J. 2004. "Music, Musicians, and the Brain: An Exploration of Musical Genius, 2004 Presidential Address." *J. Neurosurg* 101:895–903.

Post, F. 1994. "Creativity and Psychopathology: A Study of 291 World Famous Men." *British Journal of Psychiatry* 165:23–24.

Richardson, R. D. 1995. *Emerson: The Mind on Fire.* Berkeley, California: University of California Press.

Rihmer, Z., X. Gonda, and A. Rihmer. 2006. "Creativity and Mental Illness (In Hungarian)." *Paychiatr Hung* 21(4):288–294.

Rochenberg, A. and C. R. Hausman, eds. 1976. *The Creative Question.* Durham, North Carolina: Duke University Press.

Rolland, R. 1911. *Tolstoy.* New York: E. .P. Dutton & Company.

Rosner, S. and L. Edwin, eds. 1974. *Essays in Creativity.* Croton-On-Hudson, New York: North River Press.

Rothenberg, A. 2005. "Family Background and Genius II: Novel Laureates in Science." *Canadian J Psychiatry* 50(14).

___ 1990. "Creativity and Madness. New Findings and Old Stereotypes." Baltimore, Maryland: Johns Hopkins University Press.

Rothenberg, A. and G. Wyshak. 2004. "Family Background and Genius." *Can J Psychiatry* 49:185–191.

Runco, M. A. 2004. "Creativity." *Annual Review of Psychology* 55: 657–687.

Russel, P. 1929. *Emerson, the Wisest American.* Norwood Mass. Plimpion Press.

Sachs, H. 2011. *The Ninth Beethoven and the World in 1824.* New York: Random House.

Schirmeister, P. Ed. 1995. *Representative Men, Ralph Waldo Emerson (1850).* New York: Marsitio Publishers.

Shenk, J. W. 2006. *Lincoln's Melancholy. How Depression Challenged a President and Fueled His Greatness.* New York: Houghton Mifflin.

Simonton, D. K. 1997. *Genius and Creativity Selected Papers.* Greenwich, Connecticut: Ablex Publishing Corp.

___ 1999. *Origins of Genius, Darwinian Perspective on Creativity.* New York: Oxford University Press.

Simonton, D. K. and A. V. Song. 2009. "Eminence, IQ Physical and Mental Health, an Achievement Domain: Cox's 282 Geniuses Revisited." *Psychol. Sci.* 20(4):429–434.

Solomon, M. 1977. *Beethoven*. New York: K Schirmer Books, a division of Macmillan.

Sporns, O., G. Tononi, and G. M. Edelman. 2000. "Connectivity and Complexity: The Relationship between Neuroanatomy and Brain Dynamics." *Neural Netw* 13(8–9):909–922.

Steptoe, A. (ed.). 1998. *Genius of the Mind. Studies of Creativity and Temperament*. New York: Oxford University Press.

Szulc, T. 1998. *Chopin in Paris, The Life and Times of a Romantic Composer*. Boston Mass.: Da Capo Press; A member of the Perseus Books Group.

Thivierge, J. P. and G. F. Marcus. 2007. "The Topographic Brain: From the Neural Connectivity to Cognition." *Trends in Neuroscience* 30(6):251–259.

Tolstoy A. 1953. *Tolstoy, A Life of My Father*. Translated by Elizabeth Reynolds Hapgood. New York: Harper & Brothers Publishers.

Tolstoy, L. 1888. Sevastopol. New York: Thomas Y Crowell and Co.

Tolstoy, L. 1884. Confessions

Tolstoy, L. with Introduction by Malcolm Cowley. 1960. *Anna Karenina*. New York: Bantam Books. /Toronto, New York, London, Sydney.

Wain, J. 1974. *Samuel Johnson, A Biography*. New York: Viking Press.

Ward, T. B. 2007. "Creative Cognition as a Window on Creativity." *Methods* 42(1):28–37.

Wilson, A. N. 1988. *Tolstoy.* New York: Fawcett Columbine.

Yearsley, D. 2002. *Bach and the Meanings of Counterpoint.* Cambridge United Kingdom: Cambridge University Press.

Zamoyski, A. 1980. *Chopin. A New Biography.* Garden City, New York: Doubleday & Company.

Zeki, S. 2001. "Essays on Science and Society. Artistic Creativity and the Brain." *Science* 293(5527):51–52.

Made in the USA
Charleston, SC
29 April 2015